GLOBAL BOUNDARIES

The global political map is undergoing a process of rapid change as former states disintegrate and new states emerge. Territorial change in the form of conflict over land and maritime boundaries is inevitable but the negotiation and management of these changes threaten world peace.

Global Boundaries considers conceptual legal and geopolitical aspects of international borders and border lands. This book also presents a detailed discussion of Antarctica as a case of global territorial dispute.

World Boundaries is a unique series embracing the theory and practice of boundary delimitation and management, boundary disputes and conflict resolution, and territorial change in the new world order. Each of the five volumes – *The Middle East and North Africa, Eurasia, The Americas, Maritime Boundaries* and *Global Boundaries* – is clearly illustrated with maps and diagrams and contains regional case-studies to support thematic chapters. This series will lead to a better understanding of the means available for the patient negotiation and peaceful management of international boundaries.

Clive H. Schofield is Executive Officer of the International Boundaries Research Unit at the University of Durham.

WORLD BOUNDARIES SERIES
Edited by Gerald H. Blake
Director of the International Boundaries Research Unit
at the University of Durham

The titles in the series are:

GLOBAL BOUNDARIES
Edited by Clive H. Schofield

THE MIDDLE EAST and NORTH AFRICA
Edited by Clive H. Schofield and Richard N. Schofield

EURASIA
Edited by Carl Grundy-Warr

THE AMERICAS
Edited by Pascal Girot

MARITIME BOUNDARIES
Edited by Gerald H. Blake

GLOBAL BOUNDARIES

World Boundaries Volume 1

Edited by Clive H. Schofield

Routledge
Taylor & Francis Group

LONDON AND NEW YORK

First published 1994 by Routledge

2 Park Square, Milton Park, Abingdon, Oxon OX14 4RN
711 Third Avenue, New York, NY 10017, USA

Routledge is an imprint of the Taylor and Francis Group, an informa business

First issued in paperback 2016

Transferred to Digital Printing 2007

Copyright © 1994 International Boundaries Research Unit

British Library Cataloguing in Publication Data
A catalogue record for this book is available from the British Library

Library of Congress Cataloging in Publication Data
Has been applied for

ISBN 978-1-138-99191-0 (pbk)
ISBN 978-0-415-08838-1 (hbk)

Publisher's Note
The publisher has gone to great lengths to ensure the quality
of this reprint but points out that some imperfections in
the original may be apparent

CONTENTS

FIGURES

TABLES

NOTES ON CONTRIBUTORS

Ilidio do Amaral is the Director of the Centre for Geography, Instituto de Incestigacoa Cientifica Tropical, Lisbon, Portugal.

Anthony I. Asiwaju is Commissioner of the National Boundary Commission, Nigeria.

Francis Auburn is Deputy Director of the Centre for Commercial and Resources Law at the University of Western Australia Law School, Perth, Western Australia.

Peter J. Beck is Professor of International History at Kingston University, Kingston, UK.

Brian W. Blouet is a Professor in the Department of Geography, the College of William and Mary, Virginia, USA.

Brigadier Michael Harbottle (Rtd) is Director of the Centre for International Peacebuilding, Oxford, UK.

Oscar J. Martinez is a Professor in the Department of History, University of Arizona, Tucson, USA.

Hernan Santis Arenas is a Professor in the Institute for Geography, Pontificia Universidad Catolica of Chile, Santiago, Chile.

Stanley Waterman is a Professor in the Department of Geography, University of Haifa, Israel.

SERIES FOREWORD

The International Boundaries Research Unit (IBRU) was founded at the University of Durham in January 1989, initially funded by the generosity of Archive Research Ltd of Farnham Common. The aims of the unit are the collection, analysis and documentation of information on international land and maritime boundaries to enhance the means available for the peaceful resolution of conflict and international transboundary cooperation. IBRU is currently creating a database on international boundaries with a major grant from the Leverhulme Trust. The unit publishes a quarterly *Boundary and Security Bulletin* and a series of *Boundary and Territory Briefings*.

IBRU's first international conference was held in Durham on 14–17 September 1989 under the title of 'International Boundaries and Boundary Conflict Resolution'. The 1989 conference proceedings were published by IBRU in 1990 edited by C.E.R. Grundy-Warr: *International Boundaries and Boundary Conflict Resolution*. The theme chosen for our second conference in 1991 was 'International Boundaries: Fresh Perspectives'. The aim was to gather together international boundary specialists from a variety of disciplines and backgrounds to examine the rapidly changing political map of the world, new technical and methodological approaches to boundary delimitations, and fresh legal perspectives. Over one hundred and thirty people attended the conference from thirty states. The papers presented comprise four of the five volumes in this series (Volumes 1–3 and Volume 5). Volume 4 largely comprises proceedings of the Second International Conference on Boundaries in Iberocamerica held at San Jose, Costa Rica, 14–17 November 1990. These papers, many of which have been translated from Spanish, seemed to complement the IBRU Conference papers so well that it was decided to ask Dr Pascal Girot, who is coordinator of a major project on border regions in Central America based at CSUCA

(The Confederation of Central American Universities), to edit them for inclusion in the series. Volume 4 is thus symbolic of the practical co-operation which IBRU is developing with a number of institutions overseas whose objectives are much the same as IBRU's. The titles in the *World Boundaries* series are:

Volume 1 *Global Boundaries*
Volume 2 *The Middle East and North Africa*
Volume 3 *Eurasia*
Volume 4 *The Americas*
Volume 5 *Maritime Boundaries*

The papers presented at the 1991 IBRU conference in Durham were not specifically commissioned with a five-volume series in mind. The papers have been arranged in this way for the convenience of those who are most concerned with specific regions or themes in international boundary studies. Nevertheless the editors wish to stress the importance of seeing the collection of papers as a whole. Together they demonstrate the ongoing importance of research into international boundaries on land and sea, how they are delimited, how they can be made to function peacefully, and perhaps above all how they change through time. If there is a single message from this impressive collection of papers it is perhaps that boundary and territorial changes are to be expected, and that there are many ways of managing these changes without resort to violence. Gatherings of specialists such as those at Durham in July 1991 and at San Jose in November 1990 can contribute much to our under-standing of the means available for the peaceful management of inter-national boundaries. We commend these volumes as being worthy of serious attention not just from the growing international community of border scholars, but from decision-makers who have the power to choose between patient negotiation and conflict over questions of terri-torial delimitation.

Gerald H. Blake
Director of IBRU
Durham, January 1993

PREFACE

Gerald H. Blake

For better or worse, international boundaries are global geopolitical phenomena which affect the lives of millions of people, and they are among the top preoccupations of governments and the military. As the point of contact between states, land boundaries are often the focus of political strains and stresses, whether or not they are the direct cause. Boundaries also create landscapes, and fundamentally affect communications, settlement patterns, and access to resources. In states with long and insecure boundaries substantial sums are spent on their protection and management. Since the end of the FirstWorld War there has been a growing interest in the delimitation of maritime boundaries, and in cases of dispute several states have been willing to spend large sums of money in litigation to secure the best results for themselves. By 1993 approximately one-third of the world's potential maritime boundaries had been formally agreed. At the present rate of progress it may take several more decades before the offshore political map is complete. Maritime boundaries are thus destined to play a significant role in international relations for some time to come. Although land boundary delimitation began long before maritime boundaries, the process is by no means complete, in spite of the impression given by the brightly-coloured wall maps and globes familiar to most people. Indeed, one of the recurrent themes of this volume is that international boundaries do not become permanent fixtures for all time. They not only change location as the size and shape of states evolves, but they also change in their functions and in their effects through time.

Altogether, in mid-1993 there are approximately three hundred and eleven land boundaries in the world, although the precise number quoted depends on whether micro-territories are included, and whether boundaries with more than one sector are counted once or twice. There will eventually be far more agreed maritime boundaries than land boundaries, even if the number of states in the world continues to pro-liferate. Meanwhile, a surprisingly large number of land and maritime boundaries are in dispute. In the case of land boundaries the dispute is often to do with local detail over the position of the boundary to within a few metres. In others, the alignment of the boundary has yet to be agreed, and in a few cases vast tracts of land or seabed are in serious dispute. Approximately one hundred and sixty land boundaries reach the coast where they become the starting point from which offshore boundaries are delimited. When there is uncertainty over the land boundary at the coast, there can be no maritime delimitation. Similarly where there are rival claims to islands, maritime boundary delimitation cannot proceed. These issues are amply illustrated in cases discussed in the other volumes of this series which explore the processes of boundary delimitation and problems of conflict along boundaries. This volume concentrates on aspects of international boundaries which are less frequently discussed in the literature – how they function, why they change and evolve, and alternatives to the nation-state model as a way of organizing political space.

Appropriately, Chapter 1 by Oscar Martinez offers a theoretical overview of the way in which boundaries function, drawing upon the author's research along the USA–Mexico boundary. Not all borders separate peoples; in many cases border people enjoy a greater degree of interaction with each other across the boundary than they do with their national neighbours in what may be called a 'borderland milieu'. Sometimes border regions develop interests which are at odds with central government or mainstream culture. Where cross-border relations are well developed they are classed as *integrated borderlands*, as can be seen in Western Europe. At the other extreme are *alienated borderlands* where interchange is practically non-existent. Between these extremes are *co-existent* and *interdependent* borderlands. Each of these types of border displays differing degrees of transnationalism which increases as neighbouring states begin to allow free social and economic interchange at the border. Martinez concludes that borderlanders worldwide have much in common, both in the hazards to which they may be exposed, and the opportunities presented by intense cultural and economic inter-action. Borderlands clearly offer rich research opportunities for social scientists.

Anthony Asiwaju's chapter on African borderlands takes up some of these themes with a plea for African planners to begin the process of transforming borderland regions from barriers to bridges, following the pattern of Western Europe. Asiwaju argues that borderland regions are potentially vital units of regional planning at international level, but African boundaries have inherited the mutual jealousies and tensions of their European creators. European governments traditionally feared transborder cooperation in the belief that the sovereignty of the nation-state would be compromised, so that some border regions took their own initiatives towards regional transboundary cooperation. The potential for such cooperation was recognized by the Council of Europe's Outline Convention (1979) in which a dozen states pledge themselves to foster transfrontier cooperation. Similar opportunities exist in Africa for the transformation of borderlands; in some respects the opportunities are probably greater. Before this can be achieved, however, African borderlands need to be better understood, and the attitude of bureaucratic élites towards border regions will have to change.

Such visions of peaceful international cooperation across frontiers do not apply where there is serious dispute over the boundary. Michael Harbottle discusses how such disputes can be resolved by a three-dimensional approach involving peacemaking, peacekeeping, and peacebuilding. In a detailed analysis of the experience of peacekeeping operations, the author emphasizes the importance of peacebuilding activities through day-to-day interaction and joint project work in which ordinary people can be involved. The best long-term guarantee of international security might be through the creation of sub-regional cooperative associations. Peacebuilding seeks to remove the structural causes of conflict, whether social, ethnic, economic, or environmental. Until these are resolved, peacemaking and peacekeeping give only temporary respite, particularly in borderland regions. Nevertheless all three approaches are of increasing importance in the changing geo-political world today. There have been so many recent changes to the world political map that some media commentators have been led to conclude that the old framework of nation-states is about to collapse, with a number of individual states disintegrating.

Two authors offer a timely reminder that boundary changes are nothing new and need not be feared. Stanley Waterman suggests that we become attached to the maps with which we grew up, and we feel threatened by change, but change is the norm in the international system. New states will appear and others will wither away. At the same time there is clearly tension between governments, international agencies, and organizations which wish to preserve the status quo, and demands for territorial change based primarily on self-determination. The communications revolution may make countries more willing to

PREFACE

accept political patterns in which multinational states or confederations of states increasingly emerge, alongside a new generation of prosperous micro-states in strategic locations. Waterman asks whether the period 1989–91 will one day be viewed as comparable to the years 1815, 1918–21, and 1945 in the evolution of the world political map. The answer is that it probably will.

Brian Blouet's chapter on Alfred Mahan is particularly interesting in this context because it summarizes some of the views of one of the world's most influential geopolitical thinkers before the First World War. At that time Mahan thought his world was facing the imminent dissolution or readjustment of political units. He believed in the right of states to grow through a process of 'natural selection'. Mahan's views were clearly imperialistic and his predictions were sometimes inaccurate, but they offer an insight into the geopolitical thinking of another age at a time of great political flux.

Ilidio do Amaral discusses the future of the nation-state, which was the political ideal of the European world during the first half of this century, and now seems to be seriously in question. Contradictory trends can be observed, however, whose final outcome has yet to be seen. The Basques in Europe for example call for autonomy, while the European Community is working towards a new European cultural awareness, and common economic organization. This kind of ambiguity can also be seen in other parts of the world. Like other authors in this volume Amaral also sees the need for research into the changing concept of the boundary as contact and communications tend to increase, but exclusive national jurisdiction remains: 'national boundaries can be compared to the membrane of a cell as they are both separating and permeable' (page 21). There can be little doubt that the conflicting interests of self-determination and the preservation of national territorial integrity, and of national autonomy versus multinational political and economic groupings of states will become dominant geopolitical themes of the next few years.

Three chapters in this volume are devoted to Antarctica which was an obvious theme for the Durham conference in 1991 because June 1991 marked the 30th anniversary of the Antarctic Treaty System (ATS) when the Treaty Parties agreed to review its operation. The continent itself is important enough, representing 10 per cent of the world's land mass. Hitherto its main export has been scientific information, although Antarctic waters already yield valuable living resources, and there is a growing tourist activity based primarily on cruise ships. Inventories of mineral resources alleged to be located in the continent may have been rather exaggerated; in any case it was agreed in 1991 to postpone exploitation of Antarctic minerals for fifty years, much to the relief of conservationists. Antarctica is also of great importance in international relations because it represents an attempt by the international community to adopt alternatives to the partitioning of territory between states. The Antarctic Treaty has been used as a model by Mark Valencia to suggest a possible regime to solve the Spratly Island dispute, turning the region into one of international cooperation with shared exploitation of resources.[1]

Peter Beck traces how the Antarctic Treaty System has worked. The System embraces both the Treaty itself, which offers a framework designed to ensure Antarctica's status as a zone of peace, a continent for science, and a special conservation area, together with a number of associated international legal instruments. In effect, Antarctica is governed by a regime which operates through annual Consultative Meetings held by rotation in member states. Territorial claims have been shelved, and the claim lines shown on maps and atlases are largely irrelevant for all practical purposes.

The authors of all three chapters on Antarctica agree that the ATS has worked well. The greatest concern now appears to be the protection of the environment from pollution. There are also legal problems associated with the Antarctic Treaty, discussed in detail by Francis Auburn. In particular he shows that Antarctica is a continent without a domestic legal system applicable to criminal cases, and no system in civil law. It seems unlikely that these can be introduced because of the relationship between jurisdiction and sovereignty. The Antarctic contributors also agree that the ATS is likely to continue indefinitely in spite of attempts by some UN member states to replace it by a UN-based regime. The Antarctic Treaty now represents over 70 per cent of the world's population, so it has a high level of support. Hernan Santis Arenas, however, sounds a note of caution which observers outside South America might do well to heed. He sees a number of tensions on the horizon which could imply future territorial disputes and resource conflicts in Antarctica. As long as the Antarctic Treaty remains in place there is unlikely to be a serious dispute, but deeply-rooted notions of territoriality in South American society might surface strongly if the juridical status of Antarctica were to change. Argentinian geopolitical literature has taken up the 'coastal front' theory which allocates Antarctic territory to Brazil, Ecuador, Peru, Chile, Argentina and Uruguay on the basis of their extreme meridians extending through Antarctica. In the meantime the ATS is welcome evidence that there are alternatives to territorial conflict operating successfully in the modern world. Several more examples are discussed in other volumes in this series, providing glimpses of a world in which bounded territory diminishes in importance.

NOTE

1 Mark Valencia, *Malaysia and the Law of the Sea*, Institute of Strategic and International Studies, Kuala Lumpur, 1991, Annex 1, pp. 139–50.

ACKNOWLEDGEMENTS

Much of the initial work on these proceedings was undertaken by IBRU's executive officer Carl Grundy-Warr before his appointment to the National University of Singapore early in 1992. It has taken a team of editors to complete the task he began so well. Elizabeth Pearson and Margaret Bell assisted in the preparation of the manuscripts for several of these volumes, and we acknowledge their considerable contribution. John Dewdney came to our rescue when difficult editorial work had to be done. In addition, many people assisted with the organization of the 1991 conference, especially my colleagues Carl Grundy-Warr, Greg Englefield, Clive Schofield, Ewan Anderson, William Hildesley, Michael Ridge, Chang Kin Noi and Yongqiang Zong. Their hard work is gratefully acknowledged. We are most grateful to Tristan Palmer and his colleagues at Routledge for their patience and assistance in publishing these proceedings, and to Arthur Corner and his colleagues in the Cartography Unit, Department of Geography, University of Durham for redrawing most of the maps.

Gerald H. Blake,
Director, IBRU

xviii

1

THE DYNAMICS OF BORDER INTERACTION
New approaches to border analysis
Oscar J. Martinez

INTRODUCTION

Recent tendencies in various parts of the world towards increased inter-dependence and integration among nations have greatly enhanced interaction among borderlands populations. Transboundary trade, tourism, migration, and attendant social and cultural relationships, have linked regions of adjoining countries ever closer to one another.

This chapter presents three new approaches to the study of border-lands interaction. First, such interaction is viewed in terms of four models that take into account environmental and human conditions that promote or inhibit cross-boundary ties. Second, a typology of borderlands society is offered as a means of determining the degree of transnationalism among different sectors of the population. Third, the concept of 'borderlands milieu' is developed to identify characteristics that make border people unique (Martínez 1994).

MODELS OF BORDERLANDS INTERACTION

Conditions in borderlands worldwide vary considerably because of profound differences in the size of nation-states, their political relation-ships, their levels of development, and their ethnic, cultural, and linguistic configurations. Despite this heterogeneity, however, it is possible to generalize about features common to all and to posit a classi-fication scheme based on cross-border contact. As the world has evolved geopolitically, more and more borderlands have tended towards con-vergence rather than divergence, but unfavourable conditions in many areas still keep neighbouring borderlanders in a state of limited inter-

1

action. Thus in categorizing borderlands it is essential to assess cross-border movement and the forces that produce it. With such considerations in mind, four paradigms of borderlands interaction are proposed: alienated borderlands, co-existent borderlands, interdependent borderlands, and integrated Borderlands (Figures 1.1–1.4).[1]

Alienated borderlands

This model refers to borderlands where day-to-day, routine cross-boundary interchange is practically non-existent owing to extremely unfavourable conditions (Figure 1.1). Warfare, political disputes, intense nationalism, ideological animosity, religious enmity, cultural dissimilarity, and ethnic rivalry constitute major causes of such alienation. International strife leads to militarization and the establishment of rigid controls over cross-border traffic.

To say the least, such a tension-filled climate seriously interferes with the efforts of local populations to lead normal lives. International trade and substantive people-to-people contact are very difficult if not impossible. The ever-present possibility of large-scale violence keeps these unstable areas sparsely populated and underdeveloped. Borderlands which have gone through this stage in the past include the Scottish–British frontier in the fifteenth and sixteenth centuries, and the USA–Mexico border during most of the nineteenth century. Currently, alienated borderlands are found in the Middle East, Africa, Asia, and Eastern Europe.

Co-existent borderlands

Co-existence arises between adjoining borderlands when their respective nation-states reduce extant international border-related conflicts to a manageable level or, in cases where unfavourable internal conditions in one or both countries preclude binational cooperation, when such problems are resolved to the degree that minimal border stability can prevail (Figure 1.2).

A scenario that reflects evolution from a state of alienation to one of co-existence is when a serious dispute is resolved by two nation-states to the extent that international relations are possible, but not to the point of allowing for significant cross-border interaction. In effect, economic and social development that normally would take place in the region under more favourable circumstances is put 'on hold'. For example, after prolonged strife two nations may reach a general agreement

2

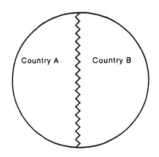

Figure 1.1 Alienated borderlands
Tension prevails. Border is functionally
closed, and cross-border interaction is
totally or nearly totally absent.
Residents of each country act as
strangers to each other.

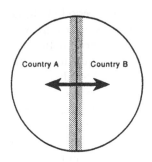

Figure 1.2 Co-existent
borderlands
Stability is an on and off proposition.
Border remains slightly open, allowing
for the development of limited
binational interaction. Residents of each
country deal with each other as casual
acquaintances, but borderlanders
develop closer relationships.

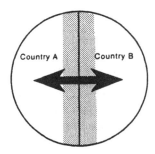

Figure 1.3 Interdependent
borderlands
Stability prevails most of the time.
Economic and social complementarity
prompt increased cross-border
interaction, leading to expansion of
borderlands. Borderlanders carry on
friendly and cooperative relationships.

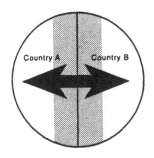

Figure 1.4 Integrated borderlands
Stability is strong and permanent.
Economies of both countries are
functionally merged and there is
unrestricted movement of people and
goods across the boundary.
Borderlanders perceive themselves as
members of one social system.

regarding the location of their common border, but leave unresolved
questions of ownership of valuable natural resources in strategic border
locales.

Another explanation for a condition of borderlands co-existence is
simply the need for traditionally antagonistic neighbours to have
enough time to get over the acrimony produced by conflicts endured

during the period of alienation. Fresh wounds, suspicion, and distrust can only be overcome with the passage of time. Eventually the elimination of overt conflict allows for enough accommodation to stabilize the border, permitting borderlanders to interact with their counterparts across the boundary within the formal parameters established by the two nation-states.

An example of a domestic issue that limits the ability of borderlanders to interact with foreigners is regional fragmentation. National unity demands a certain degree of integration among disparate territorial entities; in its absence central governments will not risk 'drift' at the periphery by allowing borderlanders to carry on substantive links with citizens of another country. In time, geographical sectionalism may be lessened through the spread of modern transportation, communication, and trade networks, diminishing the isolation of peripheries and giving the centre strong control over independent-minded frontiersmen. Once sufficient internal unity has been achieved, central governments will be less concerned with 'leakages' at the border or with contacts between frontiersmen and foreigners. Co-existence characterizes the Ecuador–Peru, Israel–Jordan, and USSR–China borderlands, to cite some examples.

Interdependent borderlands

A condition of borderlands interdependence exists when a border region in one nation is symbiotically linked with the border region of an adjoining country (Figure 1.3). Such interdependence is made possible by relatively stable international relations and by the existence of a favourable economic climate that permits borderlanders on both sides of the line to stimulate growth and development that are tied to foreign capital, markets, and labour. The greater the flow of economic and human resources across the border, the more the two economies will be structurally bonded to each other. The end result will be the creation of a mutually beneficial economic system.

Interdependence implies that two more or less equal partners willingly agree to contribute and extract from their relationship in approximately equal amounts. Of course this is an ideal state that may approach reality in some instances, but the prevalent pattern in bi-national regions throughout the world has been one of asymmetrical interdependence, where one nation is stronger than its neighbour and consequently plays the dominant role. In the case of two substantially unequal economies, the productive capacity of the wealthier countries is often matched with the raw materials and cheap labour in the poorer nation to create complementarity which, while asymmetrical in nature, none the less yields proportional benefits to each side.

Economic interdependence creates many opportunities for borderlanders to establish social relationships across the boundary as well, allowing for significant transculturation to take place. Thus the binational economic system produced by symbiosis spawns a binational social and cultural system.

The degree of interdependence in the borderlands is contingent upon policies pertaining to the national interests of the two neighbours. Concerns over immigration, trade competition, smuggling, and ethnic nationalism compel the central governments carefully to monitor the border, keeping it open only to the extent that it serves the agenda of the nation-state.

Conditions in the USA–Mexico borderlands constitute a good example of strong asymmetrical interdependence. Better balanced interdependence may be found in parts of Western Europe, where economic inequality among neighbouring nations is less of a problem than in the western hemisphere or other continents.

Integrated borderlands

At this stage neighbouring nations eliminate all major political differences between them and existing barriers to trade and human movement across their mutual boundary (Figure 1.4). Borderlanders merge economically, with capital, products, and labour flowing from one side to the other without serious restrictions. Nationalism gives way to a new internationalist ideology that emphasizes peaceful relations and improvements in the quality of life of people in both nations through trade and diffusion of technology. Each nation willingly relinquishes its sovereignty to a significant degree for the sake of achieving mutual progress.

Integration between two closely allied nations is most conducive when both are politically stable, militarily secure, and economically strong. Ideally the level of development is similar in both societies, and the resulting relationship is a relatively equal one. Population pressures are non-existent in either nation, and neither side feels threatened by heavy immigration across their open border.

Lack of data makes it difficult to cite examples of integrated borderlands, but if any region in the world reflects such conditions among select adjoining nations, it would surely be Western Europe.

BORDERLANDS SOCIETY

People in border regions are frequently closely associated with foreigners, particularly in cases of intense cross-boundary interaction. Powerful international forces tend to pull many borderlanders into the orbit of adjoining countries, with a resulting array of transnational relationships and lifestyles. On the other hand, some sectors of the population manage to remain shielded from transnational activities, and their lives are minimally affected by proximity to borders.

In accordance with these opposing patterns, borderlanders may be divided into two general types: (1) national borderlanders, and (2) transnational borderlanders (Figure 1.5).

National borderlanders are people who, while subject to foreign economic and cultural influences, have low-level or superficial contact with the opposite side of the border owing to their indifference to their next-door neighbours or their unwillingness or inability to function in any substantive way in another society. Transnational borderlanders, by contrast, are individuals who maintain significant ties with the neighbouring nation; they seek to overcome obstacles that impede such contact and they take advantage of every opportunity to visit, shop, work, study, or live intermittently on the 'other side'. Thus their lifestyles strongly reflect foreign influences. For some transnational borderlanders, such influences are modest, but for those who are seriously immersed in transborder interaction, foreign links govern central parts of their lives.

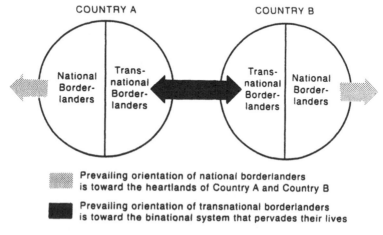

Figure 1.5 National and transnational borderlands

6

National and transnational borderlanders may be further subdivided into distinct subgroups depending on local circumstances such as the ethnic configuration and degree of transborder contact.

Along the USA–Mexico border, for example, three major population groups – Mexicans, Mexican Americans, and Anglo-Americans – interact in an environment of intense transnationalism, and the result is a wide array of border 'types' (Figure 1.6). Among Mexicans, I have identified four 'national' subgroups and five 'transnational' subgroups; among Mexican Americans, two 'national' subgroups and six 'transnational' subgroups, and among Anglo-Americans, three 'national' subgroups and four 'transnational' subgroups (Martínez 1994). While some of these subgroups are unique to the USA–Mexico border, there are certain types that are universally found in many other border areas where substantial cross-boundary interaction takes place. Such types

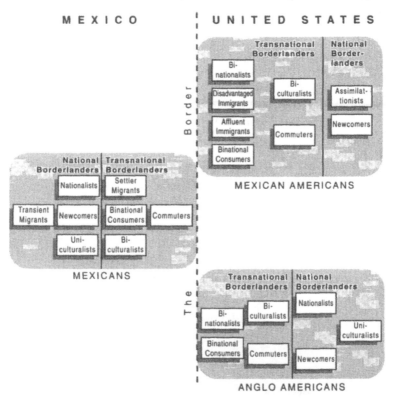

Figure 1.6 National and transnational borderlands along the USA–Mexico border

include 'uniculturalists', 'biculturalists', 'binationalists', 'worker commuters', 'binational consumers', and 'immigrants'.

Typologies of border populations are helpful for understanding relationships among different groups within the confines of nation-states and for identifying transboundary patterns. How people relate to one another reveals much about their immediate surroundings and the policies followed by nation-states in shaping that environment. The situation at the USA–Mexico border offers an excellent starting point for studying border subgroups, but comparative research with other borderlands is needed.

THE BORDERLANDS MILIEU

Whether relatively closed or relatively open, border zones are distinct within their respective nation-states because of their location, which in many cases is far from the core, and because of the international climate produced by adjacency to another country. The unique forces, processes, and characteristics that set apart borderlands from interior zones include transnationalism, international conflict and accommodation, ethnic conflict and accommodation, otherness, and separateness. In their totality, these elements constitute what might be called the 'borderlands milieu' (Figure 1.7).

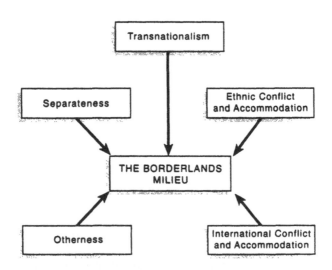

Figure 1.7 Major contributors to the borderlands milieu

Transnationalism

Location at the edges of nation-states places borderlanders in international environments that have wide-ranging implications for those who function in or are affected by transborder interchange. This applies most directly where circumstances permit extensive cross-border movement, as in the case of interdependent and integrated borderlands. Relatively unimpaired interaction makes it possible for residents of contiguous borderlands to be active participants in transnational economic and social systems that foster substantive trade, tourism, migration, information flow, cultural and educational exchanges and sundry personal relationships.

An open international environment exposes borderlanders to foreign values, ideas, customs, traditions, institutions, tastes, and behaviour. Borderlanders find it easy to see how members of other societies make their living, how they cope with daily life, how they acquire their education, and how they exercise their responsibilities as citizens. Consumers are able to purchase desired foreign products, business persons find it possible to expand their clientele beyond the boundary, and employers with a need for labour have access to foreign workers.

Transnationalism thrives when border barriers dissipate, but even under less than favourable circumstances cross-boundary interaction can still occur because total isolation from the outside world has become a practical impossibility. Fundamentally the level of transborder contact is dependent on the relationship carried on by adjoining nation-states, the concentration of population at the border, and the condition of the binational economy. In alienated borderlands, where the official policy is to maintain a closed border, movement from nation to nation may range from non-existent (in extreme cases) to slight. Normally only borderlanders who have a compelling need to carry on ties with their neighbours are willing to risk the dangers of clandestine activity. Transnationalism increases as contiguous nation-states overcome mutual antagonisms and begin to allow substantive social and economic interchange at their common border. The variation in the level of interaction under different circumstances is summarized in Table 1.1.

International conflict and accommodation

Viewed within the context of frictions that beset constituent units of nation-states, border-related strife is distinctive because it emanates from conditions peculiar to peripheries subject to international disputes

9

Table 1.1 Variations in the level of interaction in borderlands

Types of borderlands	Degree of transnationalism
Alienated borderlands	None to low
Co-existent borderlands	Low to moderate
Interdependent borderlands	Moderate to high
Integrated borderlands	High to very high

as well as border instability. Borderlanders face special challenges innate to the boundary itself, while interior populations, who live at a distance from the border environment, are shielded from such stresses.

Border people who are caught up in territorial struggles between their antagonistic nation-states will be subject to attack from foreigners or even from some of their countrymen who might question their loyalty if they express a desire to remain neutral or simply to be left alone. Often fighting goes on over long periods of time due to disagreement over the location of the border between rival countries, turning border-landareas into battlefields. Such a dangerous situation forces many borderlanders to choose between remaining in their war-torn homeland or abandoning it for safer ground.

Boundaries established by warfare or by power politics often elicit challenges from the losing side, which may entertain hopes of recovering lost territories. Or, to the chagrin of an already dismembered nation, its stronger neighbour may actually seek to acquire yet more territory. In either case one important effect is the creation of great uncertainty and instability in the borderlands, where violence remains a strong possibility as long as the border remains unsettled.

As troubled borders achieve some stability through accommodation, the dangerous climate to which borderlanders are exposed diminishes, but conflict still remains a constant feature. Bitterness and distrust produced by previous tumultuous eras may long linger in the memory of borderlanders, making it difficult to achieve significant cross-border cooperation and interchange. Any border incident, even if minor, has the potential to rekindle old hatreds and to produce confrontations.

Beyond the residual effect of historical alienation, borderlanders must confront new challenges posed by changing border conditions. Economically dynamic borderlands that have progressed to the stage of interdependence, for example, may face frictions associated with international trade, smuggling, undocumented migration, heavy cross-border traffic, and international pollution. Thus while the emergence of interdependent borderlands has diminished traditional strife related to border locations, it has not eliminated conflict. New disputes have been spawned by the intrinsic contradiction of maintaining border restrictions while the economies and societies of both sides draw closer together.

When a binational borderland reaches the stage of integration, it is possible that borderlanders may be able to shed much of the burden of being front-line receptors of international conflict and controversy. Nevertheless, even in this ideal state sources of friction will still be present, because fundamental questions about economic advantage, nationality, assimilation, and identity will continue to cause distress. In short, conditions on the border may change, but the innateness of the boundary as an agent of abrasion will not disappear unless the border itself completely disappears.

Ethnic conflict and accommodation

In contrast to populations in the heartlands of nation-states, where cultural homogeneity is the norm, people of border regions are more likely to live in heterogeneous environments owing to ethnic mixture and immigration from contiguous countries. Cultural diversity inevitably produces inter-ethnic frictions, especially if the groups represented have a history of adversity. The greater the divergences in race, religion, customs, values, and level of economic development, the more pronounced the inter-group tensions.

Ethnic strife often originates with attempts of mainstream societies forcefully to assimilate all groups within the nation-state, precipitating strong resistance among peoples determined to preserve their different identities and lifestyles. Such opposition is found especially in cases where a group is more strongly tied to people of the same ethnic background who reside across the boundary than it is to the dominant society in its own country. For the nation-state concerned with national integration, such a situation poses particularly troublesome problems. For border minorities caught up in cultural tugs-of-war, the perplexities are equally pronounced.

In the case of relatively isolated folk-cultures, discord with other groups may arise out of fear and resentment triggered by encroachment from outsiders. Settlement of unwelcome 'aliens' in the homeland of the local group in particular has the potential to unleash passionate confrontation. Such polarization has been recorded in the oral traditions and ballads of numerous conquered peoples.

11

While this discussion emphasizes the conflictual nature of ethnicity in a borderlands environment, it is also true that under appropriate circumstances harmony can be the prevailing pattern. Co-existence, accommodation, and mutual assimilation are other possibilities for different groups in close contact with one another. Such a peaceful climate offers tremendous possibilities for advancing the human condition, even if on a small scale and in isolated settings.

Otherness

Aware of the unique environment that shapes their lives, borderlanders think of themselves as different from people of interior zones and outsiders perceive them that way as well. One distinction entails relationships with citizens of other nations. Remoteness from the heartland and sustained interaction with foreigners tend to dilute nationalism among borderlanders and make them more tolerant of ethnic and cultural differences.

Another mode of behaviour observable among borderlanders that makes them different is their tendency to bend or ignore national laws deemed injurious to their regional interest. Statutes dictated by the nation-state that do not interfere with cross-border symbiotic relationships find no objection, but those that do are routinely circumvented and violated. For example, borderlanders find it morally and culturally acceptable to breach trade and immigration regulations that interfere with the 'natural order' of cross-border interaction. A popular justification given for defying the state in this manner is that such laws are unrealistic; that they fail to take into account the unique conditions in binational settings where interdependence is a way of life. Such laws, borderlanders point out, are made by distant, insensitive, and excessively nationalistic politicians. Moreover, the sentiments of borderlanders, whose usually sparse numbers and remoteness from the centre of power limit their political clout, are seldom taken into account by decision-makers.

Many borderlanders in effect live and function in several different worlds: the world of their national culture, the world of the border environment, the world of their ethnic group if members of a minority population, and the world of the foreign culture on the other side of the boundary. Considerable versatility is required to be an active participant in each of these universes, including the ability to be multilingual and multicultural. By contrast, individuals from interior zones who live in homogeneous environments have no need to develop such multifaceted proficiencies, or to be knowledgeable and sensitive to the perspectives of other peoples. Thus geographic location, economic interaction with foreigners, and cultural diversity make the lives of borderlanders stand out from the national norm.

Separateness

By virtue of their distance and isolation from the cores of nation-states, coupled with unique local ethnic and economic characteristics, border frontiers frequently develop interests that clash with central governments or with mainstream cultures. Consequently a sense of separateness and even alienation from the heartland is not uncommon in many borderlands. In his perceptive analysis of the dynamics of frontier and border areas, K.D. Kristof writes that 'frontiersmen tend to insist on a degree of detachment, autonomy, and differentness which sets them apart and is incompatible with the state interests and the rule of law as seen by the central government' (Kristof 1959). Borderlanders in particular clamour for recognition of their special needs by the authorities, often insisting that some national laws have detrimental regional effects and must therefore be changed or enforced differently in border zones.

The tendency towards 'drift' away from parent societies commonly found in frontiers is specially true in borderlands because the latter are susceptible to influences and ties to foreigners. Owen Lattimore, in his class study of Asian frontiers, concludes that:

> there grows up a nexus of border interest which resents and works against the central interest. This phenomenon of border society, differing in orientation from the bulk of the nation, recurs in history at all times and in many places. (Lattimore 1940 and 1988: 244)

The transitional nature of borderlands produces integrative and assimilative forces that blur differences between peoples on opposite sides of the boundary, spawning problems with parent populations. In extreme cases, groups on the outer edge of a nation's territorial domain often do not feel bound by rules formulated by officials situated in far-away places. Many merchants and traders in particular come to think of themselves as members of a self-contained and self-directed border economic community rather than as 'pure' citizens of a nation-state whose behaviour must conform strictly to national norms. This sense of economic independence emboldens them to circumvent laws they dislike and to carry on relationships with foreign neighbours that promote their self-interest and that of their binational region. Frontiers-men with substantial cross-boundary links often function as a 'joint community' and 'become a "we" group to whom others of their own nationality, and especially the authorities, are "they". [Thus] it is not surprising that the ambivalent loyalties of frontier peoples are often conspicuous and historically important' (Lattimore 1968: 374).

CONCLUSION

As the world has evolved from isolationism toward integration, border-lands have become increasingly important for nation-states with signifi-cant cross-border interlinks. This is particularly true in Western Europe and North America, where a number of binational borderlands thrive from pronounced trade, migration, cultural interchange, and social interaction.

Regardless of location, borderlands around the world are alike in a number of ways. As peripheries of nation-states they are subject to frontier forces and international influences. Most borderlanders are exposed to processes that have the potential for generating conflict, including border-related disputes, oppressive tariffs, restrictive migration policies, constraints to free cross-border movement, ethnic frictions, and stereotyping by outsiders. On the other hand, to be a borderlander is to have opportunities unavailable to people from heartland areas. Border-landers live in a binational milieu and are exposed to different ideas and cultures; they also have access to a foreign economy, which increases employment possibilities and consumer choices.

For social scientists, the human environment found in border areas offers truly exciting possibilities for research. This chapter has outlined general conditions that characterize the functioning of borderlands, implicitly suggesting different approaches to the study of borderlands populations. Those of us engaged in such research in different parts of the world would benefit greatly from knowing about and possibly co-ordinating our common efforts. The Durham conference in 1991 offered us an excellent opportunity to begin that process.

NOTES

1 My conceptualization has benefited from ideas on borderlands interaction posited by Raimondo Strassoldo and C.S. Momoh. Strassoldo has briefly sketched three models, 'Nation-Building', 'Co-existence', and 'Integration', while Momoh has outlined the functioning of what he calls 'Zero Border-

lands', 'Minimum Borderlands', and 'Maximum Borderlands' (Strassoldo 1989: 389–90 and Momoh 1989: 51–61).

REFERENCES

Lattimore, O.D. (1940; 1988) *Inner Asian Frontiers of China*, Oxford: Oxford University Press.

—— (1968) 'The frontier in history', in R.A. Manners and D. Kaplaw (eds) *Theory in Anthropology: A Sourcebook*, Chicago: Aldine.

Kristof, K.D. (1959) 'The nature of frontiers and boundaries', *Annals* (of the Association of American Geographers) 49: 269–82.

Martínez, O.J. (1994) *Life and Society in the US–Mexico Borderlands*, Tucson: University of Arizona Press.

Momoh, C.S. (1989) 'A critique of borderland theories', in A.I. Asiwaju and P.O. Adeniyi (eds) *Borderlands in Africa: A Multidisciplinary and Comparative Focus on Nigeria and West Africa*, Lagos: University of Lagos Press.

Stassoldo, R. (1989) 'Border studies: the state of the art in Europe', in A.I. Asiwaju and P.O. Adeniyi (eds) *Borderlands in Africa: A Multidisciplinary and Comparative Focus on Nigeria and West Africa*, Lagos: University of Lagos Press.

2

NEW REFLECTIONS ON THE THEME OF INTERNATIONAL BOUNDARIES[1]

Ilidio do Amaral

With the exception of the inhospitable polar regions and a few areas where colonial relationships survive, our planet is now covered by a patchwork of politically organized units, i.e. modern sovereign states, bounded within and confined to their legal limits: the international boundaries. However, changes are occurring.

The reality is a complex of entities, from the vast macro-states (over 8 million sq. km – the former Soviet Union/Commonwealth of Independent States encompasses 22.4 million, Canada 10, China 9.6, the USA 9.4, Brazil 8.5, etc.) to the smallest micro-states (less than 1 thousand sq. km – the Vatican, within the city of Rome, has only 0.44 sq. km, Nauru, a point in the Micronesian archipelago, 22, etc.); from the greater superpowers, rich and heavily industrialized, to the least developed and very poor states; from the most compact territorial units with high levels of accessibility and connectivity, to the extremely fragmented archipelagical countries constituted from a multitude of islands separated by international waters; from the states with long elaborated and profoundly rooted *raison d'être* to those that are trying hard to define it. However, notwithstanding such differences, invested with their sovereignty over their politically bounded territories they all have a seat in the international forum of the United Nations. But we are in a time of changing processes affecting the concepts of state and international boundaries.

Over 90 per cent of the world's independent states have some form of pluralistic structure. Peoples have attachments to several levels of territorialism, but the actual dominant structure is the state and so, as a generalization all the peoples on earth are defined by state levels within their boundaries. Unfortunately this statement is based more on theory

16

than reality for many groups do not identify themselves with the states in which they live. Some of them belong to groups split by international boundaries which do not conform to their cultural distribution. Others are minorities seeking either a degree of political autonomy within the existing states, or total separation. Beneath the surface are always problems related to the actual meaning of international boundaries.

> [First], imagine the removal of all existing international political boundaries and then, thinking of all distinctive large population groups and regionalisms, impose new political boundaries around these populations. Two things would stand out: the location of the new boundaries would be quite different from the currently existing ones and there would be hundreds of new political units ... [accepting that] hundreds of new 'states' might represent the salvation of the world! ... would a world of only small (many of them mini) states necessarily be good? Diversity is one of the key stimulants to successful societies, not uniformity. The dilemma to be faced in any such redrafting of international boundaries would be, what scale of population is 'proper' to enjoy a territory and government of its own? (Knight 1983)

According to the same author, 'there is no easy answer to this question'. Nevertheless we have to pursue a satisfactory solution for the problems related to the international boundaries.

Even in Western Europe, the home of the ideal homogeneous nation-state, ethnic divisions are readily apparent in regional and nationalist political activity. Conflicts whose origins stem from the multi-ethnic compositions of the state are the most difficult with which governments have to contend, and their severity can be great enough to threaten the territorial integrity of the state. Therefore, the ideas about the role of ethnicity need to be reconsidered in a new light. By making ethnic differences the salient cleavage the politics of plural societies are effectively defined on lines that cut across economic class divisions.

According to Rabushka and Shepsle (1972), there are four possible situations of inter-ethnic behaviour within a state: dominant majority; balanced competition; dominant minority; and fragmentation (i.e. varied network of tribe-nations). Whereas this model relates only to events within a state, there is a need to expand it in order to cover areas surpassing one state, such as two states, or a state and a neighbouring region where a minority in one state regards itself as a part of a majority in another separate but closely related territory. Two different situations can be envisaged: the minority within a state is concentrated in a small

17

area where it can be considered a majority; and the minority is distributed randomly over the state area. Different behaviour will result from these two cases, and so we can add a fifth situation to those described by Rabushka and Shepsle, when their first three situations are exhibited simultaneously, as a result of variable definitions of territory only, for example Arabs and Jews in Israel (Soffer 1983).

In the last century Europeans have been responsible for drawing the boundaries in much of the Third World and for twice redrawing the map of Europe. Since 1945 the majority of the states have been confronted with enormous difficulties related to the maintenance of their unity and even their existence as states, due to the strong emergence of regionalistic or nationalistic movements. The most remarkable feature of many contemporary separatist movements in advanced industrial states is that both types of separatism, the territorial (resting primarily on the spatial distinctiveness of the potentially independent unit) and the ethnic (resting on the cultural distinctiveness of the community pressing for independence), are being increasingly combined to produce ethno-regional (or ethno-national) movements which seek to liberate their respective peoples firmly settled in distinct, if subordinate territories.

Let us remember that the important political model that equates the cultural phenomenon of a nation (a sociological concept) with the institution of the sovereign state (a juridical and political concept) brought together in a single territory has been the most influential spatial theory in the modern world. It became the central facet of geographical contributions to the boundary drawing of European political units, especially after the First World War. Afterwards the theory has come to be neglected in Political Geography as somehow 'old fashioned'. However, the majority of the states in the world are post-1945 in origin and the geography of their state-formation and nation-building is still to be fully understood. Definitions of 'nation' vary according to language, discipline of study, and author. Quite clearly, however, a nation is a cultural identity, separated from other nations nearby. There is the tendency for aspiring nations to put emphasis on the concreteness of defined territories against the vagueness of cultural distributions. With a clearly-defined territory and an internalized cultural unity, a sense of nation may grow stronger. Certainly, this unity of group and land requires full political expression. The national-state this way becomes the political–territorial synthesis of the nation and the state. There is also the question of whether such Western European concepts as 'nation' and 'nationalism', formulated to the post-colonial

18

situations, should be transferred without change, especially in the sparsely populated areas where new states are hardly viable. Zartman (1979) gave a new dimension to the problem, stating the following about the Saharians:

> The basic political fact of the Sahara, often forgotten in the heat of higher politics, is the fierce independence of its inhabitants and their habit of self-government which favours no master. 'Independent', 'self-sufficient', 'proud': all refer to political traits born out of desert existence. The result may be admirable; it is scarcely endearing or productive of a broader political stability. It also means that integration of such populations within a nation-building state is difficult.

There are sufficient contentious points leading to the statement to merit a review of geographical question of the nation-state.

Political scientists are proclaiming the end of the nation-state as a viable entity. During the first half of the twentieth century the nation-state became the political ideal of the European world, and many political researchers assumed that true nationhood required full sovereignty. On the other hand, most multicultural states of the world (and there are many of them) have refused to accept this severely disruptive ideological position, and have preferred autonomy, or less, to the granting of sovereignty. Most definitions presuppose that the nation exists at only one level, and that level is an ultimate, superseded only by concepts such as humanity. This is probably the reason why there is a strong feeling that a nation should also be a state, and vice versa. In fact we normally allude to relations beyond the state level as being international, not interstate, and the most representative world forum refers to itself as the United Nations, and not the United States, which would be more accurate, though confusing.

In several European countries, and many of the new states of Africa and Asia, the cultural nations or their equivalent are generally allowed to develop their cultural identities, but at the same time are expected to conform to the unifying regime. This conformity is based upon the possibility of a sense of kinship to some larger social entity beyond that of the more strict cultural nation. Actually we have the opportunity to be spectators and actors of diverse political and social processes. The extension of the concept of nation beyond its strict uni-levelled application, to a dual-levelled application is a fact. In this century, a great experiment took place in former Yugoslavia, where the Croats, Serbs, Slovenes and Macedonians, though still keeping some of their national identities, were nevertheless also supposed to adopt a sense of Yugoslavian nationhood.

The new African states offer many experiments of state-formation and nation-building. All of them have adopted the model of the nation-state and the ideas of nationalism diffused by the Europeans. In order to implement the development of the model they strive to transform rapidly the pluricultural societies within the boundaries drawn by the colonial powers at the end of the last century in such uniquenesses as, for example, one Nigeria one Nigerian, one Mozambique one Mozambican, one Zaire one Zairean, and so on. The aspiring nation-states voted to build the coveted national unity upon the existing ethnical diversity, and besides that within the frame of a desired pan-Africanism. Hence the voiced clamouring against the maintenance of the traditional indigenous institutions, which they consider to be dangerous and anachronistic and accusing the tribalism, the region-alism, and the racism as being the bitter enemies of national-state building.

The Organization of African Unity has defended on several occasions the primacy of pan-Africanism and has condemned the attempts for either secessionism (within a state), or integration (of part of a state into another). The boundaries inherited at independence should be conserved and changes made only by bilateral or multicultural consent (De Blij 1991; Brownlie 1979). And the 'problem of integration is essentially one of getting people to shift loyalty from a structure based on tradition to a new artificial entity, the nation-state, whose only justification for authority lies in its constitution' (Wallerstein 1961). The majority of those new independent states are like giant puzzles of cultural pieces without a shared common language, nor the sense of common history. Being linguistically diverse, they are obliged to retain as official, the language of the ancient colonizer which functions as vehicular language or *lingua franca* to transmit the governmental decisions, the plans and interpretations of the new order. The experi-ence has demonstrated how difficult it is to select one indigenous language among the diversity because none of them has such funda-mental registers as dictionaries and grammars, and without such tools they will remain local or regional spoken languages.

In Europe, events since 1918 (when Central Europe was shattered into several states) have proved the bankruptcy of the idea of every ethnic nation forming its own state. Today Western Europe appears to be moving even further towards the national concept on three levels. While small groups such as Basques and others press for autonomy on a local scale, the established state organizations continue, and beyond that a common European consciousness seems to be growing slowly. One may point out that the EC is working steadily towards a shared consciousness of common European cultural background, towards a common economic organization, a common planning for the future, for example. In time these will surely produce a new sense of national kinship, beyond the range of the present state.

Even before the Second World War political geographers had developed a particular interest in the study of international boundaries, in spite of the prime concern being the regional differentiation of the earth's surface. The approaches have been historical, juridical and geopolitical, to clarify problems around the delimitation and demarcation, the conflicts, and the evolution of boundaries. The renewed contemporary aim is to understand better how the dynamic forces at the periphery of a state and at the interface between adjacent states impact the human development of a common borderland. Due to the unusual problems associated with borderland research – restrictions on movement, non-comparable data, language differences, difficulty of access to needed information, and the like – these regions present a real opportunity for international cooperation in research not only among neighbouring geographers but also among other specialists. Frontiers (as zonal components, and therefore containing various geographical features, including populations) and boundaries (loosely described as linear because, in fact, they occur where the vertical interfaces between state sovereignty intersect the surface of the earth) are two fundamental geographical aspects. As the sovereign state has replaced earlier forms of large political regions, it has become essential that sovereignty should have a known exact extent, a territory under exclusive jurisdiction limited by state boundaries. They are not only a line of demarcation between legal systems but also a surface of contact of territorial power structures, and their position may become an index to the power relations of the contending forces. The concept of a boundary as a separating line has been substituted by that of a contact zone with sufficient permeability. National boundaries can be compared to the membrane of a cell as they are both separating and permeable. Yet, even when the co-existence of many state ideas and 'creeds' is generally accepted (with exceptions, of course) it is important to maintain the spheres of the several integrating forces legally delimited.

The traditional boundary is losing its functions in favour of new ones, less apparent, from linear to zonal limits, from strict physical to cultural interpretations, from only spatial to functional bounds, from horizontal to vertical dimensions, from non-permeable to permeable interfaces, from administrative to social milieu. The concept of boundary is changing, with a tendency to be nearer the idea of contact and communication than that of separation and limit. *D barriera a cerniera* (from barrier to junction) is an excellent legend created by Strassoldo (1973).

Today, observers can follow the depolitization of the traditional political boundaries and the growing importance of regional boundaries both within states and at supranational levels in cases of regional integration. Proximity to the state boundary can be expected to affect human activities and landscapes in a variety of ways. The economic, sociological and psychological characteristics of the border zone are much more significant, though less obvious than the physical structures.

NOTE

1 This paper was produced in June 1991, prior to the Soviet and Yugoslav political earthquakes.

REFERENCES

Brownlie, I. (1979) *African Boundaries: A Legal and Diplomatic Encyclopaedia*, London: Hurst.
De Blij, H.J. (1991) 'Africa's geomosaic under stress', *Journal of Geography* 90, 1: 2–9 .
Knight, D.B. (1983) 'The dilemma of nations in a rigid state structured world', in N. Kliot and S. Waterman (eds) *Pluralism and Political Geography: People, Territory and State*, London: Croom Helm.
Rabushka, A. and Shepsle, K.A. (1972) *Politics in Plural Societies*, Columbus: Charles Merril.
Soffer, A. (1983) 'The changing situation of majority and minority and its spatial expression. The case of the Arab minority in Israel', in N. Kliot and S. Waterman (eds) *Pluralism and Political Geography: People, Territory and State*, London: Croom Helm.
Strassoldo, R. (1973) 'Regional development and national defence : a conflict of values and power in a frontier', *Boundaries and Regions: Explorations in the Growth and Peace Potential of the Peripheries*, Trieste: 387–416.
Wallerstein, I. (1961) *Africa: the Politics of Independence*, New York: Random House.
Zartman, I.W. (1979) 'Boundaries and nations', *Focus*, 29, 2: 2–3 and 6–7.

3

BOUNDARIES AND THE CHANGING WORLD POLITICAL ORDER[1]

Stanley Waterman

'Alle Menschen werden Bruder,
Wo dein sanfter Flugel weilt'

The years 1815, 1918–21, and 1945 are years which stand out as focal points in the evolution of the world political map, at least from the European viewpoint. It may be that when historians and others in related disciplines in the centuries ahead come to view the years 1989–91, they may perceive another significant political watershed in the development of world politics. But is the period of the early 1990s destined to be another of the important periods of realignment of the world political map? In other words, will this period prove to be of the same interest to political geographers as it is proving to be to students of international relations and political science?

Most of the changes that have occurred since 1989 have concerned Europe, a continent that was among the most stable in the preceding four decades. One of the periodic major realignments in international relations has occurred, perhaps sooner than expected. As a result of a host of internal and other difficulties close to home – economic, political, ethnic, ideological and so on – the Soviet Union either elected or was forced to relinquish its hold over Eastern Europe.

So far, there has been barely any change in the status of international boundaries. The only boundary alteration to accompany the startling political developments in the past two years has been the abolition of the 'internal' German frontier, the line which divided the two German states from the second half of the 1940s until 1990. The abolition of this border, marking the unification of the divided German nation, actually followed similar moves elsewhere in the past decade and a half, as in Vietnam, Yemen, and more recently *de facto* between Syria and Lebanon. This amalgamation of the two German republics happened as part of the process of the sloughing off of four decades of Communist rule in Eastern and Central Europe, a process apparently so spontaneous and virulent that it took everyone by surprise. In a paper written in 1985, I seemed to think that the two Germanys would eventually come to accept one another and the then *status quo* the longer they remained separated (Waterman 1987). Even such a well-informed newspaper as *The Economist* did not perceive what was to occur in Germany until it happened. Neither, it appears, did the Germans themselves.

Although democratic elections have been held in Hungary, Czechoslovakia, Romania, Bulgaria, Albania, Poland and in the new Germany, as well as in several Soviet republics including Russia, with the German exception, no boundary changes resulted from these events. Slovenia's attempt to secede from the Yugoslavian federation in the summer of 1991 might appear to have some hope of success from the vantage point of early autumn 1991, mainly because Central European states support it and the minorities problem in Slovenia is minimal. On the other hand, the desire of the Croatians to free themselves from Serbian dominance is much more complicated, not only because of the intermingling of the two but also because of their historic relationships and the corrections to their joint boundary which will prove necessary should the divorce take place. In the Soviet Union, the freely elected governments of the Baltic states, especially Lithuania, have attempted to make operational those parts of their election platforms which included a declaration of independence; the Caucasian and some Central Asian republics may follow suit; the new Union charter/constitution laying out the relationships between the central federal government and the republics currently being touted for the reformed Soviet Union does not envisage boundary changes. Attempts to alter internal Soviet borders in recent years, as in Nagorno-Karabakh, have failed.[2]

How do we accept changes in the world political map? Many of us, as individuals, but not necessarily as scholars, seem to be attached to the map with which we grew up. It becomes our norm for the world. We need a baseline from which to measure and evaluate all change.

For example, my first atlas was a *Times Atlas* in my parents' home, and I must have traced all the maps from this book as a child. It must have been published around 1949 or 1950, as Germany was divided, Poland was 'normal', Czechoslovakia was attenuated in the east, and so on. Much of Africa was still pink (British), blue (French), green (Portuguese), yellow (Spanish), or brown (Belgian), and I have always been uneasy with the Africa that emerged in the 1960s and 1970s. It was just easier with the wide swathes of French North Africa, French Equatorial Africa, British East Africa, the Belgian Congo (as distinct from the French Congo), and such like. On the other hand, my father, born in 1913 and whose impressionable years were in the decade following the First World War, was uncomfortable with the European section of the atlas: Poland was the wrong shape; where had the Danzig corridor gone, and so on.

There is another interesting situation in Israel. There, the geography of the country has always been taught in schools as the geography of the *Land of Israel*. (The Land of Israel, as distinct from the State of Israel, equates with the borders of British Mandate Palestine, i.e. Israel and the West Bank.) This, and the little truth (perhaps startling to some) that Israel has controlled the West Bank and the Gaza Strip for longer than the 'Green Line' divided Israel from the Jordanian and Egyptian controlled parts of Palestine, makes Israeli schoolchildren, students, and general citizens ignorant of what the legitimate map really is. The constant shifting of boundary lines since 1948 has greatly confused the Israeli public's understanding of what constitutes the real borders of the state, if not the ultimate ones (Bar-Gal 1979; 1991). Whereas the norm for many Israelis who entered their formative years in the 1950s and 1960s includes the Green Line and the partition of the Land of Israel between Arabs and Jews, many younger and older people fail to make this distinction. For many, the borders of the Land of Israel form the normative boundaries of the country.

We have to be blind or ignorant (or both) not to note changes such as these for what they are – the evolution of the political map. Similar political changes have occurred, piecemeal, all over the world throughout recorded history, and with increased frequency in the twentieth century. Some changes have resulted from far-reaching international agreements on how the world map was to change, *grosso modo*; others resulted from bilateral arrangements. The return of Trieste to Italy, the border agreements between Saudi Arabia and Kuwait, between Saudi Arabia and Jordan, between Nigeria and Cameroon and the partition of Ireland are all bilateral examples. The post-First World War political map of Europe created at American behest after the collapse of the great empires resulting in the emergence of new states like Czechoslovakia, Poland, Hungary, Austria; the transfer of territories from one to another, such as the southern Tyrol to Italy from Austria; of Alsace-Lorraine and the Saarland to France, of Danzig to the Poles; the division of the Middle East into new states such as Palestine and (later) Transjordan, Lebanon, Syria and Iraq, represent component parts of wider agreements.

Change in the international system is the norm. It has been, as F.S. Northedge reminded us, continuous, and never more so than in the twentieth century. Because, more often than not, change occurs piecemeal, we do not normally pay it too much attention. As Northedge put it: 'We are rarely conscious of the bones in our body until they start to ache' (Northedge 1976: 67).

The political scientist George Quester has written about the problem of incomplete surrenders following wars and the new arrangements for peace that follow (Quester 1991). The very concept of an incomplete surrender and the unwillingness of many states to 'play the game' means that *ipso facto* they are thinking not about the present but about the future. The present is not fully compatible with the future and is only a transient phase meaning that the situation after the war might have to be accommodated for a period until an improvement is registered and longer-term plans can be met.

Similarly, when Saul Cohen writes extensively about the geopolitical alignment of states (and most people are familiar with his concept of 'shatterbelts', borderlands lying on the peripheries of the great geopolitical regions which are occupied by numerous small states, each vying with the others for dominance), he sees a developmental sequence from shatterbelt to subordinate or peripheral part of a geopolitical region to a new kind of intermediate or 'gateway region' (Cohen 1963; 1973; 1991). The 'audacity' of suggesting the emergence of more than forty new states, including such concoctions as Lithuania, Slovenia, Mount Lebanon, Gaza, Kashmir, but also Western Australia, Alaska, British Columbia and Scotland foresees a major realignment of states. Such realignments would undoubtedly be painful to the states themselves, in the very process of becoming states, as well as to those states from which they have arisen. However, note that some of what Cohen was writing about less than two years ago seems less far-fetched than it did then. For example, there has been a revolution in Ethiopia in which the Eritreans came out on top; Slovenia has seceded from Yugoslavia, and the Baltic States from the Soviet Union; Hong Kong's status will change in 1997; Quebec is threatening yet another referendum on separation/independence. However, even Cohen seems not to be writing about a drawing of the world political map *de novo*, but about a hiving off of problem areas, a series of secessions, the formulation of several new little states rather than a breaking away and a coalescence of smaller units into new, perhaps larger, wholes, with new names

26

which might make more sense regionally.

It is fascinating how 1945 has formed the baseline for so many phenomena and events in the modern world, just as the 'Great War' was the baseline for our parents' and grandparents' generations. 'Since the war' and 'in the post-war era' are catchphrases that we have grown up with, as if there had not been other wars since. And this even in Israel!

Since the end of the Second World War, the world has been divided into four principal groups of states, a political organization which came into being in the late 1940s. At the end of the war, the spoils of Europe were divided between the victors – the western allies and the Soviet Union – and in geostrategic terms, the Soviets had the upper hand. Western Europe was returned to the rule of the major nation-states which inhabited it, whereas Eastern and parts of Central Europe were dominated by the military might and the political totalitarianism of the Soviet Union. In the two and a half decades following, Asia and Africa were liberated from their colonial yokes, and if they did not seek refuge in the arms of Mother Russia, they often did seek solace there.

The main groups of states are as follows:

(i) The 'West', or the 'First World', those states which have been loosely grouped around the leadership of the United States of America.

(ii) The 'Eastern Bloc States' or the 'Second World', grouped around the leadership first of the Soviet Union, and then divided between it and China.

(iii) The 'Third World', more or less synonymous with the Less-[Un]-[Non]-Developed Countries, (LDCs, UDCs, NDCs) or even the non-Aligned States, a large group of states most of which are characterized by a recent history of colonial (usually European) domination, under the leadership of India, Yugoslavia, or Cuba. The attitude of these states has been generally anti-Western and they were nurtured by the Second World states as such.

(iv) The 'Fourth World', which consists of a small group of pariah states, including South Africa, Israel and Taiwan. These states are characterized by market economies but have blemishes in their social and political behaviours. Pariahs or not, most of the rest of the world has cynically succeeded in maintaining healthy commercial relations with them while maintaining diplomatic and political boycotts.

In the European realm the political and military situation was accompanied by a changing political map. Essentially, the pre-war boundaries

27

were pushed westward, Poland at the expense of Germany, while its eastern regions were digested by the Soviet Union. The Baltic republics were lost to the USSR. The Soviet Union also claimed and acquired parts of Finland (Karelia became part of Russia, giving Leningrad some breathing space and Finland a refugee problem), Romania (Moldavia became a Soviet republic although its population was ethnically Romanian), and Czechoslovakia (the blue-eyed darling of Europe during the first round of twentieth century political reorganization after the Great War, found that its eastern tail had been snapped off and appended to Ukraine). Most startling of all was the partition of Germany, the most powerful state in Europe into the Federal Republic and the Democratic Republic, with the ludicrous isolation of the former and soon-to-be-again capital, Berlin, within the sovereign territory of the Soviet-controlled Democratic Republic of Germany.

These political–geographical events were accompanied by demo-graphic movements on a vast scale. The Second World War saw an enormous number of wanton deaths – Jews and gypsies and many others in the German death camps; 10 million Soviet soldiers and citizens killed directly by hostilities, starvation and Stalin. Moreover, there were enormous population shifts, as people were moved around to fill the political–territorial units that it was thought most aptly suited them. Germans were moved out of Poland, the Soviet Union, Czecho-slovakia and elsewhere into Germany; East Germans, unhappy with the state established there fled in their millions to West Germany which accepted them and absorbed them as full citizens, to its clear benefit. There was a general 'cleansing' of the nation-states of East and Central Europe, where there had been shifts in the boundaries (Kosinski 1969). This was not so in other cases, such as in Yugoslavia, where Serbs and Croats, Slovenes and Albanians, Bosnians and others, continued to pester and compete with one another for autonomy and power in the artificial federal state which had been concocted out of that most balkanized of all regions.

The world political boundaries that emerged at the end of the Second World War were based on social and economic ideological differences rather than differences based on the perception of cultural and ethnic variability. The ideology of the Communist world negated the legiti-macy of ethnic and cultural differences while at the same time acknowledging their existence. The ideology of the capitalist West did not negate this variety, rather it tried to mitigate it. The Western world could live with the ideological division of Europe and the rest of the world so long as it did not interfere with its economic growth and advancement.

The Communist world's acknowledgement of national differences – not only in the Soviet Union and Yugoslavia – through the use of internal passports and identity papers, and by the inability to replace linguistic and other cultural markers with the ideology of the Marxist– Leninist–Maoist brands of socialism (Smith, 1991) meant that these variations lay dormant throughout the forty years of Communist hegemony in those parts of the world which had fallen for or to its charms. The underlying differences of hundreds of years lay waiting to be unleashed as the overlayer of forty years of ideological variety was removed. The mitigation of the national question in the non-Communist world to the benefit of economic advancement only partially hid similar conflicts. The language question between Walloons and Flemings in Belgium; Basque and Catalan demands in Spain; Protestants and Catholics in Northern Ireland are all very much part and parcel of the emerging United Federation of Europe that may or may not be in the offing.

The European pattern was repeated in parts of Asia, although there appears to have been a concatenation of events that had occurred in Europe following both the World Wars. In the Indian subcontinent, tens of millions fled as British India gave birth to the modern Indian and Pakistani states, and again some two decades later when the two-headed Pakistani state devolved into its component parts, Pakistan and Bangladesh. What had been French Indochina was partitioned into first four (two Vietnams, Laos and Cambodia), later into three units; Korea became two in much the same way as Germany, and there were two Chinese states each claiming to be the one and only. The federation of Malaysia was created, resulting shortly after its creation in the separation of Singapore and the evolution of a new, small, powerful and unique city-state.

History may be the story of man and his civilizations, the story of changing times and changing fortunes. To the geographer interested in history, it represents what has occurred in particular places over given time. Most geographers interested in history call themselves historical geographers, and the more the pity because historical geographers have tended to concentrate on a given place at a given time such as England in 1086, pre-Famine Ireland, or the Holy Land in the nineteenth century. They have left the historians to write the great sweeping stories of man's intellectual and political development.

Some political geographers of the past were interested in such great things. I think that Isaiah Bowman's great post-First World War survey (Bowman 1919 and 1924) was one such attempt to look at the world as

29

a whole. Saul Cohen's books (Cohen 1963; 1973) and later papers (Cohen 1982; 1983; 1991) represent a similar global approach. Peter Taylor's recent romance with the world economy (Taylor 1988; 1989; 1990) and John O'Loughlin's (O'Loughlin 1990; O'Loughlin and Heske 1991; O'Loughlin and van der Wusten 1990) and Geoffrey Parker's with geopolitics (Parker 1985; 1988) are other examples. There are many more.

If history teaches political geography anything, it is that nothing is static. Our conservative attitude to the world political map results from the fact that the world, divided into sovereign political units, is like an apple cart, piled high with apples. Nobody wants to be the one to upset the apple cart in case his national apple should also fall, unplanned, from the cart (Waterman 1987). Not only that, of course, but some of the apples on the cart are rotten (even to the core) and should have been removed years ago, but getting at them is often a dangerous and unhealthy job.

Thus, although the world changes constantly and continuously, many people choose to ignore this and pretend that we have stability. The status quo is maintained by the stance of international organizations, which are large bureaucracies with vested interests, much concerned with 'jobs for the lads'. Certain international organizations, such as the European Community, which have a definite aim (i.e. the closer integration of European states and the mitigation of nationalisms, though state identities and representations are scrupulously maintained) are actually concerned with bringing about a simultaneous radical and conservative change. However, a more common example is the Organization of African Unity whose charter avoids tampering with existing state units, most of which are the result of the very colonial period from which the states involved struggled to free themselves. Self-determination for African states is certainly favoured, but without the dismantling of boundaries. This absurd attitude was adopted as part of the fight of African statesmen interested in the modernization of their states and in which they saw tribalism as a major barrier to the realization of their goals. Thus to have redrawn the political map would have been to surrender to the scourge of tribalism. However, consociational democracies in which groups in society agree to differ and take turns at sharing power are very difficult to establish and maintain, and do not work everywhere. They exist in countries, such as the Netherlands, where the cleavages in society are recognized, but put aside to make way for the more important business of running the state. One wonders at the chances of success of such a potentially wealthy state as South

Africa/Azania, when one looks at the inter-tribal conflicts, not just involving the White tribe against the non-White, but within the Black and White communities themselves. Who has determined, *a priori*, that the new South Africa which is emerging must remain within the boundaries of the present South Africa?

The United Nations is prepared to admit a finite number of new states (not nations) who are shaking off colonialism and imperialism. One wonders what kind of welcome the former Soviet republics (two of which, Ukraine and Byelorussia, have been full members of the General Assembly since its foundation) who choose to stay out of the reformed union, might receive if and when they 'get their act together'. Such is the UN antagonism towards changes in the world political map that the organization goes out of its way to 'resolve' all border and ethnic disputes peacefully, not necessarily for altruistic reasons alone but also to avoid upsetting the apple cart. Of course, this is only the tip of the iceberg. There are many other shortcomings of the UN which cry out for reform, such as permanent membership of the Security Council. At present this is composed of the two superpowers – the United States and the Soviet Union, the most populous state in the world – China, and two former colonial powers, one of which perhaps recognizes its new second or third rank status, while the other still believes in itself as a civilizing entity, neither of which is the largest or richest of the Western states but both of which happened to be on the winning side half a century ago. The two most successful powers in the First World were the big losers in 1945, but neither has yet received full recognition of the status that it has attained in peacetime, and both are reluctant to accept the mantle of a world power, as the 1990 Gulf Crisis and War only too vividly portrayed. The UN has no real permanent role for the LDCs, unless China is considered such. India is not a permanent Security Council member either. Moreover, the UN still plays the game in the General Assembly of every state as equals – thus China, India, the USA or Japan, to name but a few have equal status with the Cocos Islands, Tuvalu, Luxembourg, and St Kitts, but not with Liechtenstein, San Marino or Monaco. I wonder if they would be accepted if they requested membership. (See Morrill (1972) on a proposal for reforming the United Nations.)

Although the world political map must constantly alter to meet the ever-changing needs of leaders and inhabitants, change is unsettling. Stability, i.e. lack of change, is easier to live with, at least for those not directly involved in demands for change. However, this might lead us to conclude dangerously that stability, or lack of change, is healthy.

Stability can lead to stagnation and myopia. Problems are seen only in national terms and the international nature of many problems – such as the environment, transportation, trade, finance – are ignored or seen as someone else's difficulty. Openness and flexibility are principal ingredients of change, and a willingness to be open and flexible, not just to approaches to practical problems and issues but to evolving international relations, might lead to greater acceptance of the need to alter the world political map.

Today, almost every state is involved somehow in the affairs of every other state. We should remember Peter Taylor's note that sovereignty is never just a matter for a single state – other states are always involved in one state's sovereignty or desire to achieve sovereignty (Taylor 1989: 141). One of the principal functions of the sovereign state is not just the organization of what goes on inside the state, but also the ability to control what passes across its borders. Perhaps we should make the distinction, as Alan James has done, between independence which refers to the relationship between the state and other states, and sovereignty which relates to the state acting within its own territory (James 1986: 14–25).

The territorial division of the world political map is partly a result of the fact that territory is so closely associated with sovereignty, and remembering the administrative principle in Walter Christaller's organization of central places, people usually do not like sharing sovereignty. The $K = 7$ principle means drawing discrete lines around areas of influence so that each smaller settlement is subordinate to a single, larger administrative centre, and shared between higher-order centres, as in the market- and transportation-optimizing situations. Administration almost always means dividing the map into discrete units. How many instances are there of shared sovereignty – Andorra or perhaps colonial arrangements between Britain and France in the South Pacific? Just consider the difficulties concerned with the Anglo-Irish accords of 1985. Today, this is more important than ever, so the whole world has to be discretely divided, as everywhere is now within the world economy.

Part of the story of state stability and friability is related to the identity of and identification with the state – perhaps something akin to what Hartshorne termed its *raison d'être*. The more the inhabitants can identify with the reason for which the state exists, the greater the stability of the state. Whereas in the past, the state was in the persona of its ruler, today the state tends towards the collective personae of its citizens or of some dominant sector of the population. As we approach the modern period, the state began to mean much more to its citizens, if only because it was increasingly becoming the purveyor of 'public goods'. What had been the divine right of kings has become the divine right of nations. Does a state have a right to exist if its boundaries do not include a nation? The question of how we define a nation and a state has been adequately dealt with elsewhere (Smith 1991).

The years 1989–91 have been a period of change, but so far, other than the balance in the relationship between the two superpowers, what has taken place has been predominantly a European experience. Africa, or parts of it, may be following suit, but in a different way. Regimes have been or may shortly be replaced in Ethiopia, Angola, South Africa and elsewhere. So far, it is difficult to trace the beginnings of political change in Asia and the Americas, and what may happen there might also be little more than a reaction to the new chemistry of international relations (cf. Vietnam, Korea, Cuba – although the relations between the Canadian Confederation and Quebec is something different). But what we have witnessed so far has been mainly political change, in particular within individual states. There has not been much to attract the political geographer directly as a political geographer.

The active Eurasian scene is witness to parallel processes of what *The Economist* labels 'fission' and 'fusion'. *Fission* is taking place primarily in two areas. The Russian Empire was assembled in the eighteenth and nineteenth centuries and added to after the Second World War, and is the only one of the multinational empires of those times not to have collapsed by the 1950s. It is in danger of splitting into tens of separate, independent units, formerly comprising republics, autonomous regions, and so forth. The other area is Yugoslavia, the only one of the Eastern European units created in the wake of the First World War not to have solved the bulk of its minorities problem in one way or another after the Second World War. *Fusion* is what may occur in the rest of Europe should the processes set in motion in 1957 and 1987 bear fruit in 1993 and thereafter, and should the applications of Austria, Sweden and then Switzerland, Finland, Norway and others to join the EC be successful. *Perhaps* a European federation will be formed; *probably* there will be greater freedom to cross borders.

The question we should be asking as macro-political geographers is to what extent these events will bring about a realignment of states, and what shape this realignment will take? Will there be some major changes? Will these be followed by minor rectifications of interstate boundaries?

There is little doubt that the world is changing, perhaps more rapidly

33

than we think. The increased mobility of money and information, especially in the past two decades, appears to be making the nation-state less relevant to people than in the past. Political leaders may not relish this, as it is plainly easier to appeal to the emotions of French people or Germans or Canadians than to employees and dependants of Royal Dutch/Shell or General Motors. Nevertheless, the past twenty years have seen financial transactions and transfer of information – use of telephone, telex, fax, electronic mail – become virtually instantaneous and out of the hands of governments, as the use of fax machines during the Tiananmen Square student demonstrations of 1989 or the Intifada have shown. Coupled with this, the growing feeling of common problems among an increasingly well-informed (or at least widely informed) population throughout the world makes the need for international cooperation appear all the more necessary and further weakens the vitality of the nation-state, at least in those nation-states in which the underlying common factor is citizenship and civic rights rather than belonging to a specific ethnic group. All these factors seem to make countries readier to accept the idea of multinational states or at least confederations of states. The difference, of course, between the emerging multinational state of the twenty-first century and that of its pre-twentieth century predecessor is that the new state will be based on democratic principles.

NOTES

1 This chapter was written during the summer of 1991 and updated later that year. The paper was therefore composed prior to the collapse of the two great European socialist federations, the Soviet Union and Yugoslavia, and the disintegration of the Czechoslovak federation. These events engendered substantial change in status for several boundaries and provided for the creation of thousands of miles of 'new' international boundaries. The author's comments, from the vantage point of late 1991, are, however, most perceptive particularly concerning predictions on the fate of the seceding republics of former Yugoslavia.

2 Nevertheless the aftermath of the attempted coup in the USSR in August 1991 seems to indicate that boundary changes will follow rapidly after declarations of independence, as the Russian republic flexes its muscles in the process of metamorphosis from the Soviet Union.

REFERENCES

Bar-Gal, Y. (1979) 'Perception of borders in a changing territory – the case of Israel', *Journal of Geography* 78: 273–6.
—— (1991) *The Good and the Bad: 100 Years of Images in Zionist Geo-*

graphical Textbooks, Queen Mary & Westfield College: Department of Geography, Research Papers.

Bowman, I. (1919; 1924) *The New World: Problems in Political Geography*, (First & Second editions), London: Harrap.

Cohen, S.B. (1963) *Geography and Politics in a World Divided*, New York: Random Press.

—— (1973) *Geography and Politics in a World Divided*, New York: Oxford University Press, Second Edition.

—— (1982) 'A new map of global geopolitical equilibrium: a developmental approach', *Political Geography Quarterly* 2: 223–41.

—— (1983) 'American foreign policy for the Eighties', in N. Kliot and S. Waterman (eds) *Pluralism and Political Geography – People, Territory and State*, London: Croom Helm.

—— (1991) 'The emerging world map of peace', in N. Kliot and S. Waterman (eds) *The Political Geography of Conflict and Peace*, London: Belhaven Press.

James, A. (1986) *Sovereign Statehood*, London: Allen and Unwin.

Kosinski, L.A. (1969) 'Changes in the ethnic structure in East-Central Europe, 1930–1960', *Geographical Review* 59: 388–402.

Morrill, R.L. (1972) 'The geography of representation in the United Nations', *The Professional Geographer* 24: 297–301.

Murphy, A.B. (1991) 'The emerging Europe of the 1990s', *Geographical Review* 81: 1–17.

Northedge, F.S. (1976) *The International Political System*, London: Faber and Faber.

O'Loughlin, J.V. (1990) 'World power competition and local conflict in the world', in R.J. Johnston and P.J. Taylor (eds) *A World in Crisis: Geographical Perspectives*, Oxford: Blackwell.

O'Loughlin, J.V. and Heske, H. (1991) 'From "Geopolitik" to "Geopolitique": converting a discipline for war to a discipline for peace', in N. Kliot and S. Waterman (eds) *The Political Geography of Conflict and Peace*, London: Belhaven Press.

O'Loughlin, J.V. and van der Wusten, H. (1990) 'The political geography of Panregions: the theory and an empirical example of Eurafrica', *Geographical Review* 80: 1–20.

Parker, G. (1985) *Western Geopolitical Thought in the Twentieth Century*, London: Croom Helm.

—— (1988) *The Geopolitics of Domination*, London: Routledge.

Quester, G.H. (1991) 'The growing problem of incomplete surrenders; neither war nor peace and its geographic implications', in N. Kliot and S. Waterman (eds) *The Political Geography of Conflict and Peace*, London: Belhaven Press.

Smith, A.D. (1991) *National Identity*, Harmondsworth: Penguin.

Taylor, P.J. (1988) 'One worldism', *Political Geography Quarterly* 8: 211–14.

—— (1989) *Political Geography: World-Economy, Nation State and Locality*, London: Longman.

—— (1990) *Britain and the Cold War: 1945 as Geopolitical Transition*, London: Pinter.

Waterman, S. (1987) 'Partitioned states', *Political Geography Quarterly* 6: 90–106.

4

THE PROBLEM OF ASIA AND THE WORLD VIEW OF ADMIRAL MAHAN

Brian W. Blouet

Alfred Thayer Mahan is widely regarded as an advocate of seapower. Certainly his writings in naval history including *The Influence of Seapower upon History, 1660–1783* (1890) and *The Influence of Seapower upon the French Revolution and Europe 1793–1812* (1892) had an impact on naval thinking, but to view Mahan as only an advocate of seapower is to underestimate his range as a strategic thinker. Much of Mahan's writing from the beginning of the century, until his death in 1914, was concerned with continental power and the problem of Asia. In some respects Mahan can be seen as anticipating the view of Halford Mackinder in relation to Eurasian landpower.

In 1900 Mahan collected a number of articles he had written in *Harper's New Monthly Magazine* and the *North American Review* and had then published by Little, Brown as *The Problem of Asia and its Effect upon International Policies* (Mahan 1900). The book is repetitive and difficult to read besides being a great disappointment to Mahan in terms of royalties. However, embedded in *The Problem of Asia*, are many of the issues that have dominated western strategic thought in the twentieth century. The following important points are made:

(i) All the leading states including Russia, France, Germany, Britain and the United States of America were undergoing a phase of expansion and acquisition of additional territory (Mahan 1900: 4).

(ii) The USA after the war with Spain (1898) had acquired territories in the Caribbean and Pacific. As a result of controlling Hawaii and the Philippines, the USA was becoming 'an Asiatic power, with consequent responsibilities and opportunities' (Mahan 1900: 11).

(iii) The USA could not look with indifference on a change in the balance of power in Europe (Mahan 1900: 17).

(iv) The Russian empire occupied a vast, central area in Asia which projected, in wedge-shaped form, into central Asia (Mahan 1900: 24–5). Only on the flanks of the Russian empire could restraint be applied (Mahan 1900: 26).

(v) Britain, in possession of the Indian subcontinent, was in a position to contest further Russian expansion (Mahan 1900: 27–8). If competition developed between Russia and Britain it would be a struggle between land and sea power.

(vi) Transport by sea was cheaper and more flexible than by land. Russia was remote from the sea and this was an economic disadvantage. Even where Russia had access to the oceans, waterways like the Black Sea and the Baltic could be shut up by hostile powers (Mahan 1900: 43). 'Only parts of the Russian territory ... enjoy the benefits of maritime commerce. It is therefore the interest of Russia ... to reach the sea at more points and ... to acquire by possession or by control ... extensive maritime regions ...' (Mahan 1900: 44). Russia should be allowed access to the sea on the coast of China (Mahan 1900: 120) and in return the seapowers should be given greater access to the Yang-tse-Kiang to help promote both economic and political development in China. No one should be allowed to dominate China and there should be both a commercial and cultural open door (Mahan 1900: 167).

(vii) Britain and the USA 'the two great English-speaking nationalities' (Mahan 1900: 139) had common principles and ideals and, in an age when humanity was tending to aggregate into larger groupings, closer cooperations between the two countries could be expected (Mahan 1900: 141). The British navy was still an essential element in America's strategic calculations. Mahan was against a formal alliance with Britain in 1900 but he expected that Britain would reduce its Caribbean naval presence and leave that sea to the USA.

(viii) 'The Atlantic, north of the equator, is the ocean of that old community of European civilization upon which, from our point of view, the welfare of humanity rests' (Mahan 1900: 191).

(ix) France tends to lie outside the community of North Atlantic states as indicated by her alliance with Russia (Mahan 1900: 129). The three 'Teutonic nations' – Germany, Great Britain, and the USA – have many interests in common (Mahan 1900: 109, 123 and 127).

(x) Japan had already made a transformation and has a solidarity of interest with Germany, Britain and the USA (Mahan 1900: 63–4 and 148–51). Germany and Japan lay on the flanks of Russia and Mahan implies that those states, in association with Britain and American naval power, could constrain possible Russian expansion even if Russia was allied with France. In the Middle East, however, Russia and France might be able to dominate (Mahan 1900: 67). To meet this possibility Britain should take a tighter hold on Egypt and the Suez Canal (Mahan 1900: 80–2).

(xi) Mahan questioned whether 'the extensions of the Monroe Doctrine to the extent of supporting the independence of the states of extreme South America against all European interference' was wise or tenable (Mahan 1900: 201 and 135–9).

(xii) Mahan thought the world was entering upon a phase in which it was 'confronted with the imminent dissolution of one or more organisms, or with a readjustment of their parts' (Mahan 1900: 46). Within the home dominions of the European and the American powers no marked territorial changes are to be expected; in the outer world, where conditions are unsettled, we can expect changes particularly in the Pacific, East Asia, Africa and the Middle East. Mahan explicitly states that he does not believe that the indigenous peoples, in the outer world, could claim any 'natural rights' to run their affairs. Resources had to be developed and utilized for the general good. If necessary, 'compulsion from outside' should be employed (Mahan 1900: 98). The viewpoint is imperialist but it does suggest that Mahan was sensing the development of a world economy.

Not all of the points present accurate predictions. In twenty years the home dominions of the Russian, Austro-Hungarian and German empires were shattered and within half a century the overseas empires of Britain and France were dissolving. It is remarkable, however, that Mahan did discuss so many themes that became central to the international affairs of the twentieth century. *The Problem of Asia* fails to foresee important details like wars with Germany and Japan but as we approach the centenary of the publication of the book the general themes discussed in the work are still recognizable today. Above all *The Problem of Asia* has not been solved from a western perspective. The problems of the Russian empire and China are still with us.

At the beginning of the twentieth century Mahan anticipated extensive change in the power of states and the territory they controlled.

What was his view of boundaries and boundary change?

Mahan held the view that states had a 'right to grow' (Mahan 1900: 30–2). This assertion of the need to grow is not advanced as a justification for manifest destiny in North America but in terms of 'contending impulses in Asia' (Mahan 1900: 31), particularly the contest between Russia and Britain (Mahan 1900: 33) in the region.

There should be a process of 'natural selection' (Mahan 1900: 46) to allow spheres to expand and, in the case of Russia, to gain better access to the sea. Mahan thought the process could be achieved by weighing balances of strength and adjusting accordingly. He did not see this shifting of spheres and boundaries as being achieved by war. What is spelled out in terms of China is a concept of porous borders but no transfer of territory. In the context of its time Mahan is advocating informal rather than formal imperial control.

In point (xii) above is the idea that large political structures will be reshaped and that economic expansion and resource exploitation will take place in a context where the sovereignty of boundaries are de-emphasized. Mahan's view is a version of the imperialism of free trade that sees national boundaries as barriers to trade and development.

The 'right to growth' is a variant of Ratzel's '*Lebensraum*'. Mahan states explicitly that conflict over boundaries will lead to hostilities. This view is developed by Bowman in a commentary on the new Europe he had helped to create in 1919:

> The danger spots of the world have been greatly increased in number, the zones of friction lengthened. Where there were approximately 8,000 miles of old boundary, there are now 10,000 miles and of this total more than 3,000 miles represent newly located boundaries. Every additional mile of new boundary ... has increased the possible sources of trouble between unlike and, in the main, unfriendly peoples. (Bowman 1921: 3)

As we face the prospect again of the imminent dissolution of one or more organisms or a readjustment of their parts, perhaps we should bear in mind the ideas of Mahan and Bowman, writing before and after the First World War.

REFERENCES

Bowman, Isaiah (1921) *The New World*, New York: World Books.
Mahan, A.T. (1900) *The Problem of Asia and its Effect upon International Policies*, Boston: Little, Brown.

5

PEACEKEEPING AND PEACEBUILDING ALONG BORDERS

A framework for lasting peace

Michael Harbottle

The resolution of conflict and the peaceful settlement of disputes require a three dimensional approach – peacemaking, peacekeeping and peace-building. All three are required to operate concurrently and to inter-relate with each other to provide a coordinated response to the process of settling a dispute in a peaceful manner. However, each has its specific aim and objective and can only achieve results in its own particular field of operation. That said, it is the peacebuilding dimension which is pre-eminent. Peacebuilding focuses on the removal of the structural causes of conflict – the social, economic, humanitarian, ethnic, sectarian and even the environmental influences which create the situation in which open, or manifest, violence erupts. Until these conflict elements are satisfactorily resolved no amount of peacemaking and peacekeeping can have a lasting effect. They are only transitory and cannot provide a permanent settlement of a dispute in themselves. They can end the physical violence and succeed in bringing the protagonists to the confer-ence table, but so long as the underlying problems continue to motivate the actions of people there will be, inevitably, a continuing threat of violence reactivating itself. Peacemaking and peacekeeping are limited in their lasting effect; only in combination with the peacebuilding process can there be a hope of a long-term peaceful solution. Examples abound in the post-World War era where political and military inter-ventions have failed to bring about the sought-after result. Even those which may have restored the status quo (e.g. the Falkland Islands) have not solved the problem about which the war was fought in the first place.

Peacebuilding, therefore, is the prerequisite to achieving the peaceful settlement of disputes. How it can achieve this will be described later in this chapter but for the moment let it suffice to point out that the peacebuilding element is required before (the preventative phases), during and after (the restoration and rehabilitation phase) a conflict. It is an ongoing process to which the peacemaking and peacekeeping initiatives need to relate. But as the peacebuilding dimension has not yet been adopted as a part of the three dimensional process, and as peacekeeping and, possibly to a lesser degree, peacemaking are viewed with some scepticism by the more reactionary in political and military circles, it is important that their definition should be clearly understood. There is a difference between peacekeeping and peace enforcement. The former is a UN concept based on the principle of no use of force in the execution of its role to end fighting. The latter, as it clearly implies, calls for the use of all necessary military means to end a conflict. While the latter enjoys the authority of the UN Charter (Chapter 7) the former does not. Though the peacekeeping concept is one which the UN has made its own, it was not considered as a role for the UN when the Charter was drafted, and therefore has no place in it. It is an omission which it would be highly desirable to rectify.

It is worthwhile making a comparison between the two methodologies because of the sensitivity which arises wherever a territorial or border dispute occurs. If we look first at the use of peace enforcement in situations where attempts have been made to settle disputes in the traditional way, we have only to study the history of the last forty-five years to see how inconclusive in military terms these attempts have been. The Vietnam and Afghanistan wars were defeats for the USA and the Soviet Union respectively, despite the strength of the armed forces and sophistication of the weaponry deployed against inferior forces and equipment. In Korea (1950–54) a stalemate situation resulted after four years' fighting and continues to this day with 45,000 American troops still stationed in South Korea and the confrontation between the two halves of the country unresolved. In the 1950s and 1960s the so-called internal security operations were undertaken by Britain, France, the Netherlands and Portugal to frustrate the attempts of the indigenous people of their respective colonial territories to achieve their self-determination. In every case the military actions failed in their purpose and the people of the countries concerned gained their independence and are today nation-states. In Northern Ireland, after twenty-five years, the British have so far failed to achieve a solution to the sectarian problem despite a major military operation in which many hundreds of soldiers (and civilians) have been killed. Finally, the Falkland Islands and Gulf wars have illustrated that military might may bring about the defeat of opposing armed forces but leave behind a situation for which a solution still has to be found. The issue of sovereignty of the Falkland Islands has still to be settled and will probably require a UN initiative to achieve it. Kuwait may have been liberated but one cannot be sanguine about the way in which the internal situation developed with reprisals against the local Palestinians and the delayed progress towards a more liberal and democratic process in the government of the country.

In this catalogue of examples from the military record of the last forty-five years I have referred to only a handful of cases, but they should be sufficient to show that enforcement action does not always provide the basis for the peaceful settlement of a dispute; nor does it begin to resolve the underlying issues and problems which are the cause of the manifestation of the violence. Take for instance the plight of the Kurds in Iraq. The Kurds remain at the mercy of Saddam Hussein who has lost no authority or domination over the Iraqi people in spite of losing a war which was intended to dethrone him. In a previous work 'The Kurdish Connection' I pointed out that the situation in which the Kurds now find themselves stems from the encouragement given to them by President Bush and others to rise up and remove Saddam Hussein (Harbottle 1991). But when they did so there was no support forthcoming from the western powers to ensure that they were successful. I will deal with the 'Kurdish question' in more detail later but is has to be recognized that its solution is as important to a peaceful settlement of the Middle East's problems as is the issue of Palestinian self-determination.

In contrast to the peace enforcement approach to conflict resolution, the United Nations' mounting and conduct of its peacekeeping operations are quite different. Because they do not conform with normal military practices and because enforcement is no part of their operational mandate, UN methodology is criticized by the more orthodox purveyors of military doctrine as being ineffective and an invitation to a continuing status quo. Of course one does not claim that all UN operations have been successful – some have proved wholly inadequate – but if one studies the effects of the eighteen operations so far mounted they provide a more encouraging picture than the peace enforcement operations previously listed. In at least four of the major operations to date – UNEF[1]1 and 2 in the Middle East, UNFICYP[2] in Cyprus and ONUC[3] in the Congo – positive and constructive results were achieved. In all cases the presence of the UN force quietened a volatile conflict, restored a degree of stability and order, and offered the opportunity for the protagonists to come to terms with the situation and enter into a dialogue for ending the violence. The fact that a UN force does not seek to use its presence to dictate a solution but rather to assist the respective parties to reach a settlement by creating an atmosphere in which that settlement can be reached, is its strength. It is a matter of interest that, whereas the British military intervention in Cyprus in the 1950s failed to overcome the guerilla activity of EOKA (the Greek Cypriot freedom-fighting organization), despite the overwhelming superiority of the British force, UNFICYP with a third of the strength brought to an end the charged and violent intercommunal fighting in the 1960s without firing a single round of ammunition in the process of doing so. In the two Middle East operations, UNEF 1 and 2, the UN created a buffer between the opposing armies of Egypt and Israel and through physical interposition for ten years in the first instance, and six years in the second, prevented a renewal of the fighting, and at the same time guaranteed a more peaceful existence for the inhabitants in the confrontation area. In the Congo, which has posed one of the most difficult operational tasks for UN peacekeeping so far (territorially the Congo was equivalent in size to North-West Europe), ONUC provided a calming influence. The then Tunisian Foreign Minister, Mongi Slim, described its performance and presence as follows:

> What can be asserted beyond doubt is that the UN presence prevented the cold war from settling in the Congo, that the unity of the Congo was reestablished ... and that the UN helped to avoid an impending chaos that threatened peace and security, not only in the Congo, but in the whole of the African continent. (Mongi Slim 1963)

There is no standard operational approach to UN peacekeeping. Each conflict scenario calls for a different mandate and *modus operandi*. While the principles can be set, the character and conduct of every peacekeeping operation will vary to meet circumstances within which it has to be implemented and executed. UN peacekeeping, therefore follows no stereotype but has to be flexible to meet the situations which the particular conflict creates. In a nutshell, peace enforcement calls for a confrontational intervention while peacekeeping depends upon co-operation and interpositioning.

In its thirty-seven years' history the UN peacekeeping machinery has been used on numerous occasions to monitor boundary agreements. In the Middle East the UNTSO,[4] established in 1945, following the creation of the state of Israel, continues to this day to monitor the observance of the treaty provisions governing the security of the borders between Israel and its Arab neighbours. Of course, there have been hiccups over the years but by and large the actual agreements remain intact. Similar observer missions have operated along the borders of Kashmir,[5] between India and Pakistan,[6] and Iran–Iraq following the cessation of the eight years' war between those two countries.[7] In 1958 the UN Observer Group in the Lebanon monitored the Lebanon–Syria border to discourage infiltration by irregular militia from Syria, while the Golan Heights and the Israel–Syria border are still being monitored by a Disengagement Observer Force (UNDOF) after seventeen years. Though it cannot be claimed that any of these missions have brought about a peaceful settlement of the long-standing differences between the countries concerned, one can but wonder what would have been the outcome in terms of armed conflict had they not been deployed. I would hazard the guess that the bloodshed would have been considerably greater and the instability and insecurity of the boundary zones more acute. The fact that in two cases (UNTSO and UNMOGIP) the length of time the UN missions have been in position exceeds forty years testifies to the value which the countries concerned attach to the continued presence of the UN along their borders, but also to the disinclination of the countries to reach a solution to their political problems.

In a different context it can be claimed that peacekeeping forces themselves have contributed to the stability along interstate borders and to preventing the renewal of fighting across them. UNEF 1 between 1957 and 1967 in the Gaza Strip and along the Armistice line between Egypt and Israel kept the peace to such an extent that the inhabitants on either side were able to farm and move freely right up to the dividing line (for the first time in decades) without fear of molestation from the opposite side. This could have had an accumulative effect on the development of good community relationships between Arab and Jew had not President Abdul Nasser insisted on UNEF 1 being withdrawn in 1967 to allow him to prosecute the disastrous Seven Days' War which followed. Although the Suez Canal does not constitute an international boundary *de jure*, it became de facto between 1967 and 1973 after Israel occupied Sinai. When fighting broke out once more between Egypt and Israel in 1973 (Yom Kippur War) it was UNEF 2 which provided the buffer between the two armies and for six years supervised the security and humanitarian safeguards of the inhabitants within the 'war' zone. By the same token the 'green line' established between the two communities in Cyprus in 1963 could not be said to be a boundary division under international law, but it also assumed de facto significance in 1974 at the time of the Turkish invasion. In all the years that UNFICYP has been in place it has provided humanitarian and economic safeguards for those trapped in the confrontation zone.

Peacekeeping by inference is a sobering influence. Because its purpose is, in part, to encourage cooperation and to reduce tension, a peacekeeping force's involvement with the population is essential to the success of its operation. In situations, such as that in Cyprus, in the pre-1974 period, the force was dispersed throughout the island, with detachments located close to villages and rural communities. Besides ensuring the security of the community as a whole its role was to facilitate the freedom of movement of the inhabitants so that they could go about their daily business and pursue their livelihoods without interference and hindrance. When necessary it provided transport and escorts to enable members of one community or the other to attend hospital and school outside their own area. At sowing and harvest time UN soldiers would escort the farmers and their labourers into the fields where these fields were in areas occupied by the other community and under the control of the latter's military or paramilitary forces (Harbottle 1970). These acts of economic and humanitarian assistance made life bearable and helped to create confidence in the capacity of the UN to safeguard civilian rights. The whole economic assistance programme of the UNFICYP was designed to reconstitute normality for the isolated and vulnerable.

The success of this humanitarian role was due in a large part to those contingents of the UN force which depended upon conscripts and volunteer reservists for their personnel. These people came from civilian jobs – carpenters, doctors, farmhands, factory workers, garage hands, house-builders, health and hospital workers, schoolteachers, shopkeepers. They therefore enjoyed a special affinity with the civilian population and one which the professional full-time soldiers did not share. It should never be accepted that peacekeeping is a job only for the professional and not for 'amateurs'. As impartial soldiers, favouring neither one side nor another but putting the human interests of the whole population first, there is much that a peacekeeper can do to enhance the day-to-day existence of the community. It is of particular interest that it is exactly this that the Royal Marine Commandos from Britain and units from the USA, France, Belgium and Italy undertook to do for the Kurdish people in Iraq in 1991.

The Kurdish initiative was yet another case in point and different again from anything which had gone before in terms of disaster relief

45

and security provision. Operation 'Safe Haven', designed to bring the Kurdish people out of the mountains and back to their villages in the plains, was an imaginative initiative on the part of the British government. What it meant was that many thousands of Kurdish lives were saved as a result of a highly professional disaster relief operation. It was probably the personal touch which ensured the operation's success, and it is good that the military should have been the architects of what is seen to have been a most complex and sensitive operation. I believe that it should have been a matter of moral responsibility to continue the operation until a substitute UN Force could be deployed to provide the protection which the Kurds demanded and needed.

The UN's slowness in deploying a force of sufficient strength and composition to substitute for the professional forces already on the ground was disappointing. Political and financial considerations should not have been allowed to slow the process. The Kurdish issue, though *de jure* a domestic concern of Iraq, does, in my opinion, constitute a threat to international peace and security and gives the UN the right to intervene to safeguard the Kurds' safety. A force could have been raised with little delay – there is no lack of countries whose troops have a comprehensive experience of UN peacekeeping, and I understand that some of them promptly offered to make contingents available. Once the commitment is made, practice shows that there need be little delay in getting the contingents on the ground. The argument that the UN cannot involve itself or interfere in a domestic dispute (Charter: Article 2(7)) becomes null and void where international peace and security are under threat. The Kurdish issue is, as I have pointed out, a matter of international concern and as such is a threat to international peace. In this event the Security Council is empowered to act (Charter: Article 39). Therefore there should have been no handicap to the mounting of the necessary peacekeeping initiative.

It could be argued that the Kurdish issue does not represent a boundary dispute. It depends, of course, on how far back in history one is prepared to go. Kurdistan, as an autonomous region, existed until 1922 when Britain and France bowed to an obdurate Kemal Pasha of Turkey and by the Treaty of Sevres divided Kurdistan into five portions between Iraq, Persia, Syria, Turkey and the USSR. However, the Kurdish people have been fighting for a return of autonomy over their own affairs ever since. What we have therefore in Iraq today is a dispute over the right of the Kurds to a separate entity – another 'out-of-law' boundary dispute. In view of the importance of a settlement of the dispute to security and stability in the Middle East it seems appropriate to include Kurdistan in this study of 'peacekeeping and peacebuilding along borders'. It also brings me to the application of the peacebuilding process in the settlement of border and other disputes.

From the preceding discussion it is clear that any military peace-keeping intervention can involve a humanitarian and civil rights role, but only as a subsidiary involvement to the primary role and cannot be more than peripheral to the solving of the deep structural causes of conflict, confrontation and violence. Something more thorough, more extensive and far-reaching is required. I have emphasized earlier in this chapter that peacebuilding is needed before, during and after the conflict. In 1966 when the fighting and intercommunal violence in Cyprus had been brought to a conclusion, UNFICYP was asked to propose to the Security Council what its future role and composition should be. UNFICYP recommended that a civilian operation should be substituted for the military one (on the lines of that deployed in the Congo, alongside ONUC) with the object of 'rebuilding the bridges' of communication, understanding and intercommunal cooperation between the two communities. This would have entailed a coordinated operation by the UN specialized agencies. Instead of the military it was proposed that the existing UN civilian police element should be expanded to provide police supervision of civil rights. In addition it was recommended that in order to ensure against a repetition of the inter-communal violence a small rapid deployment unit of regimental strength should be included and held as a reserve for any emergency. The emphasis was on peacebuilding but as is the case with many radical ideas the proposals were rejected. It has always been to my regret that the Security Council did not see the wisdom of revising UNFICYP's mandate and adjusting the Cyprus operation to suit the new situation. Had it done so I believe strongly that it would have helped to avoid the catastrophe which overtook the island in 1974 and which created the distress and anguish which its people still experience.

The pattern of the civilian operation introduced into the Congo in 1960 could serve as a basis for the development of the peacebuilding concept. The operation was complementary to ONUC and aimed to assist the Congo towards a viable infrastructure of government. When Belgium pulled out, it left behind no trained civil service to take over executive responsibility for the running of the country. Out of a popu-lation of 40 million there was a cadre of only seventeen university graduates from which to build, plus the lower grade civil servants of clerical status. It was a remarkable achievement that in the seven years that the operation lasted, ONUC, and later UNDP (United Nations

Development Programme), built up a civil service and administrative infrastructure which was second to none in Africa at that time.

The chief weakness in the Congo civilian operation was that the specialized agencies which took part operated independently of each other in respect of their individual mandates. There was little lateral communication and liaison, each agency working through its own chain of command to its head of department in New York or Geneva. There was virtually no coordination in the field or at headquarters level with the result that the effectiveness of the operation lost some of its impact. However, in terms of rehabilitation and reorganization much was achieved in overcoming the obstacles. Clearly the problem was one of 'agency sovereignty', but it is an important procedural question to resolve and a lesson to be learnt. In any future operation of the kind all the agencies (and non-governmental organizations which might be involved) should accept a loss of sovereignty as part of a single co-ordinated command and control structure, in which they interrelate and interact with each other. Only then will the peacebuilding process operate to the best advantage.

Disaster relief is already recognized as being an ever recurring dilemma facing the human race. It has now been decided in the UN to set up a central supervisory body to oversee and coordinate future disaster relief operations on a global basis. This is a long-awaited decision by the UN and one can only hope that it follows the principles of operating described above. But of course the focus of this new body is on natural rather than manmade disasters, as a result of war. While I have emphasized the contribution which UN peacekeeping forces can make to peacebuilding in the course of their duties, the execution of Operation 'Safe Haven' in Iraq gives a very clear indication of the vital life saving role that the military can play when there is both an operational and humanitarian responsibility to fulfil. But one has to recognize that the military has defence and security duties to perform for its own country and so cannot indefinitely be deployed in situations such as in Iraq today. That is why it was necessary for a UN peacekeeping and relief operation to be mounted as quickly as possible, (i) to provide the safety shield for the safe havens and (ii) to provide the peacebuilding dimension of rehabilitation, reconstruction and reassurance of the Kurdish people, so that they could face the future with confidence and with a sense of security on which they could rely. A variation on this theme is presently being attempted in former Yugoslavia.

At the beginning of this chapter I made the point that peacebuilding has a conflict prevention dimension. So far I have dealt only with the in-conflict and post-conflict phases. Now I wish to address the preventative scope of peacebuilding which could be used in forestalling possible conflict between states. It is a general comment that the United Nations only enters the fray *after* the conflict or violence has already begun and that its powers of preventing them occurring are negligible and ineffective. This is fair criticism though the UN's diplomatic initiatives and the use of its Secretary General's 'good offices' have on more than one occasion deflected the warlike intentions of governments to more peaceable approaches for settling their differences. Boundary disputes have often been resolved simply by reaching a compromise in a redrawing of the boundary, which before had been marked differently on the maps of the respective disputants. (A map marked with a coloured chinagraph pencil looks fine until it is realized that its thickness represents 800 yards on the ground.) However, it has been the inability of the United Nations to halt the progress of war in different parts of the globe which has done most to undermine opinion as to its capacity for maintaining international peace and security. It is therefore a matter of importance that the United Nations should develop those structures which can preempt the threat of war through an ongoing process of peacebuilding. What follows outlines just one such structure which I suggest could provide a basis on which to build permanent interstate relationships of cooperation instead of the perennial confrontation which normally dominates such relationships.

Over the past three years a new entente has developed between east and west in Europe with the result that the enemy image syndrome has largely disappeared and attention is being paid to new concepts of security in Europe. If cooperation can replace confrontation in European interstate relations a parallel could be sought in other regions of the world. Regional organizations already exist in the Americas, Africa and South-East Asia, while the Arab League constitutes the Middle East regional body. But on the whole these regional bodies have proved unwieldy and largely impotent when called upon (under the provisions of the UN Charter) to sort out disputes and threats of disputes within their continental boundaries. This is not surprising when one remembers the territorial size of and the ethnic and cultural mix in the areas for which they are responsible. It therefore seems sensible to break down these geographical conglomerates into more manageable proportions, into sub-regional collective security systems, whereby small groups of states can coordinate a common policy for economic, social, cultural, environmental and military security. The holistic approach to security, covering all aspects which threaten it, is of

49

paramount importance in the achievement of global peace and security. Peace along and across borders would greatly enhance this prospect. The idea of sub-regional collective security would be to encourage the maximum cooperation, interaction and interrelationship between its members states, so that where one state is threatened economically, or by some environmental crisis, the sub-region reacts as a whole and comes to that state's assistance. In the same way, where a state might be threatened by another, within or outside the sub-region, the sub-region would make a collective response to end the threat. A precedent exists in respect of Costa Rica which has not had an army for forty-five years. It relies for the integrity of its borders on the Treaty of Friendship and Mutual Assistance, ratified by seventeen member states of the Organization of American States in 1948. Twice in that time Costa Rica has been threatened with attack from Nicaragua but on both occasions the treaty provisions have been successfully applied.

The Costa Rican experience could form the pattern for interstate safeguards within the collective security system. The provision of some mutual assistance guarantee could certainly affect the size of military budgets of individual states by reducing the present size of their armed forces and the degree of expenditure on arms and equipment, thus making more money available for the more crucial aspects of development and economic growth.

Whatever the structure devised, and it would most likely be different for each sub-region, there would have to be a willingness on the part of the members to surrender some degree of sovereignty in return for a shared cooperative within a group of states. In doing so the latter would have to be seen to provide greater security, stability and prosperity than exists under existing arrangements.

The process of creating the necessary structure does not have to start from scratch. Consultative bodies already exist on a sub-regional basis (e.g. the Southern African Development Coordination Conference and the Gulf Cooperation Council) and these could be expected to acquire a greater significance within the new structure. But at present not all of them operate a permanent secretariat but set up temporary offices to manage the periodic conferences and meetings. To ensure that the sub-regional security system does operate effectively and is able to respond quickly at any time in respect of its security, and economic and development interests, it would seem necessary for some kind of permanent secretariat to be created which would monitor and process the developing situation within the sub-region. It would also service a consultative council which would be expected to meet regularly to coordinate sub-regional policy in all its aspects.

This council would operate on a permanent footing with sitting members drawn from the highest ministerial ranks of the respective governments. The terms of reference of the council might include the following:

(i) Formulation of common security procedures for avoiding internal sub-regional conflict.

(ii) Coordination of a sub-regional collective security and defence strategy.

(iii) Liaison with regional and United Nations organizations on collective security and development measures.

(iv) Coordination of joint initiatives in the face of an external threat.

(v) Cooperation in the study of military requirements called for by (i)–(iv).

(vi) Cooperation and coordination in meeting the economic, social and development needs of the sub-region.

(vii) Collaboration in pooling respective national resources for the benefit of the sub-region as a whole.

(viii) Formulation of joint plans for combating environmental problems within the sub-region.

(ix) Development of joint marketing and purchasing enterprises.

The principle terms of reference for the council would be to develop policies and strategies aimed at providing acceptable standards of security and stability, particularly in respect of mutual assistance as this will govern the extent to which governments can reduce their respective military expenditure programmes. Even so sub-regional collective security on its own may not be considered to have sufficient viability and member states might look for stronger guarantees. These could be provided in two ways – by the Regional Organization and the United Nations.

The Regional Organization will continue to have its responsibilities for maintaining peace and security as defined in Chapter VIII of the UN Charter. So that it retains this overall responsibility, the region's Secretary General should be an ex-officio member of its sub-regional councils, so that he/she is well informed of any situation developing which calls for regional action.

In the Palme Commission's Report of 1982 a three-phase UN initiative was proposed whereby the UN could intervene with a view to assisting in the prevention stage of any conflict. The initiatives were as follows:

(i) On being alerted by at least one of the disputing parties to the danger of possible conflict, the (UN) Secretary General would constitute a fact-finding mission to advise him/her on the situation.

(ii) If circumstances warrant, and with the consent of at least one of the disputing parties, the Secretary General would seek the authorization of the Security Council to send a military observer team to assess the situation in military terms and to demonstrate the Council's serious concern.

(iii) In the light of circumstances and the report of military observers, the Security Council would authorize the induction of an appropriate UN military force at the request, or with the consent, of one or both of the disputing states with a view to preventing conflict. This force to be deployed within the likely zone of hostilities ..., thereby providing a visible deterrent to a potential aggressor. (Palme Commission 1982)

Here we have a preventive action designed to be deployed before the manifest violence erupts – a new approach for the United Nations to consider. However, even this degree of preventive action is not likely to be sufficient to satisfy the sub-region. Member states would be looking for even stronger safeguards in the form of preventive diplomacy from the United Nations and these could be forthcoming in the following manner:

(i) The appointment by the Secretary General of a permanent Special Representative, chosen for his/her experience as an international diplomat, his knowledge of the sub-region and of the languages and the customs of its states; a person known and well respected by the leaders of the countries concerned. The Special Representative's task would be to act in a diplomatic capacity to help promote the internal security and stability within the sub-region through personal relationships with the individual Heads of States. The Representative would also act as the eyes and ears of the UN Secretary General with regard to any direct or potential threat to the sub-region as a whole and to the states individually and would need the services of a small secretariat support. A precedent for such a role exists in Britain's Malcolm Macdonald's roving diplomatic mission in Southern Africa in the 1950s, while a comparable role was performed by Dr Sture Linner of Sweden who, as Head of the UN Development Programme for the Middle East, was adviser and confidante to the respective governments.

(ii) The UN Security Council could be encouraged to set up ad hoc 'good offices commissions', using approved and impartial diplomatic missions accredited to the countries of the sub-region; an action the Security Council is entitled to take under the provision of Article 29 of the Charter, whereby it is empowered to set up subsidiary organs deemed necessary for the performance of its functions. Their function would be to monitor continuously the security situation and report back to the Security Council any development which created a potential threat to the viability of the sub-region and its collective security structure, so that the Security Council could act early in preempting possible conflicts. This also has a precedent. In 1946 the Security Council, under the authority of Article 29, set up a Good Offices Commission (GOC) to monitor on its behalf the handover by the Netherlands of their East Indies' colonial territories to the Indonesians. The GOC relied on selected consular missions in Jakarta for day-to-day information and advice about the progress and management of the handover.

Such additional safeguards would provide further protection against external threats and would assist in the early settlement of disagreements and disputes before they assumed dangerous proportions (Centre for International Peacebuilding 1992).

It may be thought that the possibilities for sub-regional security systems have been over simplified, particularly in the light of the reality of violent conflict in the world as a whole. Can one hope for something of the kind evolving in the Middle East as an outcome and in the aftermath of the Gulf War? Certainly it could form the basis for a new cooperative interdependence in that region, but it will depend upon a clear intent and willingness on the part of the states of the region, including Israel, to work out the structures for peace and security and evolve a cooperative association using the Conference on Security and Cooperation in Europe (CSCE) pattern on which to build. But the structure has to be of the region's own making and must be appropriate to the economic, cultural, humanitarian and environmental needs of all the people. It should not be imposed from outside by the developed countries as a part of superpower politics, one of whose worst habits is to believe that they know best what the countries of the developing world require. They set the global geopolitical perspectives without understanding that each country, each region assesses its perceptions of security and evolution quite differently. If the Middle East is to achieve security and stability in the future it will be as a result of the states of the region developing a collective *modus operandi* relevant to their needs and their perspectives. The industrial nations can play a vital support role in the provision of assistance, expertise and technological know-how, on an as required and requested basis. They should subordinate their role and involvement to one of consultancy and unconditional economic aid.

It is so often the case that mechanisms for change exist but are ignored because of a lack of imagination and incentive. This is true of cooperative security. In 1975 the European countries, east and west, plus Canada, the USA and the USSR drew up and endorsed the Final Act of the Helsinki Accord. It contains a comprehensive blueprint for building confidence and strengthening security through interaction. Though those sections of the act dealing with military and human rights issues have been the subject of much publicity, the section which deals with economic, social, educational and environmental cooperation has only in recent years been given much attention, and then largely on the initiative of the non-governmental organizations, rather than the governments themselves. Yet that section sets out a catalogue of inter-action, exchange and joint project work in which ordinary people can be involved. It is peacebuilding in practice and demonstrates that peace is everybody's business. Real security and stability can only be achieved when there is a total involvement of the people in its achievement.

Though the Helsinki process was designed for implementation in Europe its relevance is not confined to that one continent. The pro-visions of the Final Act can easily be modified and adapted for application in other countries, regions or sub-regions. It forms a valu-able model, illustrating how confidence-building, cooperation and peacebuilding can be combined to create new concepts of collective security in the developing and developed world.

Security, to be absolute, has to take account of everyone's sense of insecurity. Each nation must recognize that there is not just one but many perspectives of security, depending upon where one lives on the planet and how one views the world from that place. If global security and stability are to be achievable, everyone needs to become aware of the different standpoints – through communication, interaction and cooperation.

Earlier, I commented that the sub-regional concept of cooperation and security is not a consideration open only to the developing and underdeveloped countries. Far from it, for circumstances now prevailing in Europe and the search for a common structure under the umbrella of the Conference on Security and Cooperation in Europe (CSCE) have created a new awareness of the continent's evolution towards a single entity and the desire on the part of Central and East European states, formerly of the Warsaw Pact, to gain membership of the European Community, along with those non-aligned and neutral countries also seeking admission. The CSCE structure could become overloaded and heavily bureaucratic. The creation therefore of sub-regional cooperative associations, as illustrated through a finlandization/balkanization process, could help to provide a stronger base from which the CSCE could develop. The idea that eventually the principle of collective security might be applied to the whole continent is still a fanciful thought, but not impossible, sometime in the future.

In a world where peace and security are measured in terms of military strengths and are conditional upon parity or superiority of weapon power, it may be difficult to visualize how security could be based on any other quantifying factor. On the other hand it is both negative and dangerous to accept the premise that weapon power is the ultimate guarantee of world peace. Peace is not just the absence of war. Peace is the establishment of international and interstate peaceful relations, understanding and cooperation. It is peace of mind and freedom from fear, threat and mistrust. However, for the developing countries particularly, security is relative to the overall world situation and has to be set and structured within it. By tradition security is a sovereign responsibility of the state and of its government. Collective security arrangements between states are normally complimentary to, not instead of, military security policies. I suggest that the time has come when we should look differently at our strategies and doctrines and devise new concepts for international peace and security – the cooperative approach instead of a confrontational one. The three dimensions of peacemaking, peacekeeping and peacebuilding are a good place to start.

NOTES

1 United Nations Emergency Force (UNEF).
2 United Nations Peacekeeping Force in Cyprus (UNFICYP).
3 Organisation des Nations Unies au Congo (ONUC).
4 United Nations Truce and Supervisory Organization (UNTSO).
5 United Nations Military Observer Group in India and Pakistan (UNMOGIP).
6 United Nations India–Pakistan Observer Mission (UNIPOM).
7 United Nations Iran–Iraq Military Observer Group (UNIIMOG).

REFERENCES

Centre for International Peacebuilding (1992) 'Collective security under sub-regional arrangements', unpublished paper presented at the conference of the Commission of the International Peace Research Association in Kyoto, Japan, July.

Harbottle, M. (1970) *The Impartial Soldier*, Oxford: Oxford University Press.

—— (1991) 'Kurdish connection'. *Boundary Bulletin* 2: 29.

Mongi Slim (1963) Hammarskjold Memorial Lecture at Colombia University.

Palme Commission Report (1982) *Common Security*, London: Pan Books.

6

BORDERS AND BORDERLANDS AS LINCHPINS FOR REGIONAL INTEGRATION IN AFRICA

Lessons of the European experience

Anthony I. Asiwaju.

Europe and its separate states must not only ensure, each at its level, that internal regions are coordinated, but also work towards cooperation with external regions, that is with regions outside the European state system. The policy of cross-border co-operation has no sense if it is limited merely to countries within the European Community which have already begun to integrate. It must be extended to other countries, too, whether or not they belong to the EEC, or the Council of Europe. (Quitin 1973: 41)

There is perhaps no other single theme that has dominated public debate about Africa's future than the implications of 'Europe 1992' for the continent. The perception of the Single European Act coming to full effect in that year has been especially sharpened by the outbreak of the democratic revolutions in Communist Europe in November 1989, leading to the virtually irreversible process, not only of the evolution of 'a Greater Europe' but also of the termination of the Cold War and the systematic dismantling of obstructionist ideological borders between the West and the East.

However, as illustrated by the extremely illuminating two-day high-powered international 'Seminar on European Community after 1992: Consequences for Africa', held in Lagos from 14–15 June 1990 under the auspices of Nigeria's Ministry of External Affairs, the debates have focused more on matters of effect than those of fundamental cause. While a great deal has been said and written on the urgency of regional integration on the model of the EEC, far less than desirable attention

57

has been paid to the more critical causal factor of the nation-state territorial structure and the border problematics.

Europe 1992 is 'Europe without frontiers'. Africa cannot respond effectively to the challenges without having directly to confront the question of the borders of the nation-states, which have given rise to the need for regional integration in the first instance. There has to be a concerted effort to convert those borders from their prevailing postures as ramparts into a new veritable function as bridges. Africa's searchlight for lessons in the European historical experience must transcend the EEC to embrace a firm grasp of knowledge about the extremely important complementary role of the Council of Europe with particular reference to the handling of the strictly territorial and border dimensions of the evolution of the European Community.

The purpose of this chapter is to draw attention to an international economic integration strategy which, while still being ignored by policy-makers and planners in Africa, has in recent decades gained in importance and achievements in Europe and is currently being actively recommended for the adoption of governments in North America. Rather than keeping rigidly to the approach whereby development is initiated continually from the top, usually through the operation of conventional sovereignty-preconditioned multinational intergovernmental organizations, the new strategy is in the effective utilization of micro- or grassroots-level initiatives and their policy potentials; it operates within the framework of the more limited intergovernmental organizations in which local or provincial authorities along international boundaries play key roles as agents of international cooperation. In this new context, the unit of action is not whole continents or their main sub-regions, as in the case of the more conventional multilateral intergovernmental organizations; it is, instead, the border[1] regions or borderlands defined characteristically as the 'sub-national areas whose economic and social life is directly and significantly affected by proximity to an international boundary' (Hansen 1981: 19). While each of such 'sub-national areas' may be planned and developed within contexts of individual nation-states, observable geographic, demographic, cultural, economic and historical links and inter-penetrations with the 'sub-national areas' on the other side of the given international boundary make the national approach less realistic and less desirable than the international transboundary approach to planning. This imperative for local planning across an international boundary has given rise to the use of a new policy instrument and tool of analysis in international relations, referred to for want of a better term as 'sub-national micro-diplomacy' (Duchacek 1986a: 11–30 and 1986b; Duchacek, Latouche and Stevenson 1988). The use of this mode has allowed for formalized cooperation among local authorities, based on pre-existing networks of informal relations between communities across the binational boundaries and covered directly or indirectly by provisions of cooperative treaties between nation-states along the specific borders.

There are obvious advantages which the use of sub-national microdiplomacy has over the alternative of the more established macro-level diplomacy. Apart from addressing itself to the issue of borders and their functions, so basic to all discussions of international cooperation, the new strategy makes for relatively easier demonstration of relevance of international economic integration efforts to the geographical and historical realities on the ground. Indeed, as has been noted, transboundary economic cooperation at intergovernmental level normally results from the formalization of the informal network of relations, which have existed for decades and, in some cases, centuries at the level of communities or culture area astride binational boundaries.

This advantage enables transboundary cooperative efforts to respond better to criticisms which generally deride macro-size intergovernmental organizations for international cooperation as something essentially remote from the people. In this regard, it is pertinent to refer to one of the series of insightful observations in a study commissioned by ECOWAS (Economic Community of West African States) and conducted by the ECA (Economic Commission for Africa) on 'Economic integration in West Africa' to the effect that 'international efforts at economic cooperation in the sub-region are, for now, largely an affair of government, and not quite of the people' (UN–ECA 1984: para 181). And if, as the study has correctly asserted, 'economic cooperation is for the people' there cannot be a more logical starting point than the borderland communities. Such communities have suffered more than their counterparts in other sub-national areas from the obstructionist effects of borders; they are, therefore, better situated to appreciate the value of transboundary socioeconomic planning and development.

Besides, borders constitute the locus for most of the interchanges between the limitrophe nation-states demarcated by them. Accordingly, borderlands should be viewed and treated as the most suitable grounds for training and testing the sincerity of nation-states and governments committed to international cooperation including efforts at regional economic integration. The relative smallness of the area and of the intergovernmental organization involved makes for greater ease of effectiveness of control and monitoring.

Finally, the use of 'sub-national micro-diplomacy, ultimately allows for continental and sub-continental levels of aggregation. In other words, there is the macro dimension of this essentially micro-level operation. This point is supported by evidence not only of the widespread nature of the phenomenon of borderlands and small nations which in Africa, as in Europe (as we shall presently examine), have the attributes of border regions.[2] Such available evidence further emphasizes the importance to be attached to the recognition of borderlands as vital units of regional planning at international levels. Some observers of the European efforts at transboundary cooperation see such cooperation as 'a significant step towards eventual European economic and political integration', and Niles Hansen, arguing with particular reference to the USA–Mexico border, has commented quite usefully that the European experience 'should prove instructive for ... other counries where common border region problems and opportunities are still neither understood nor appreciated adequately in the respective national capitals' (Hansen 1983).

Whatever our sensitivity about comparisons which make Europe the reference point for Africa, it is fair to consider the border regions on the latter continent as essentially a replica of those on the former. In Europe, as in Africa, with particular reference to the western sub-region, neighbouring border regions represent areas of distinct official language, national histories and cultures, differing economic systems and administrative organizations.

This is not surprising since the borders, which spawned the borderlands of Africa in the first instance, were creations of European imperialists who drew and, for a long time, managed them on the model of the borders of their own respective metropolitan countries. As has been discussed in several scholarly works, the borders of modern Africa are so much a European superimposition that all the legal instruments for dealing with them have remained exactly the same 'agreements', 'treaties', 'protocols' and 'notes' which the particular European powers established between or among themselves at the time of the drawing and the maintenance of the colonial boundaries (Anene 1970; MacEwan 1971; Brownlie 1979).

The rulers of independent African states which, at least in territorial structural terms, are for the most part no more than successors of the former European colonies, have to this extent maintained the status quo: not only were the legal instruments inherited, the institutions, personnel and the mode for dealing with the subject of boundaries have remained basically the same as their European antecedents or their derivatives. Little wonder then that border relations in Africa have continued to feature the same kind of mutual jealousy, conflict and tension, and have continued to be managed within the framework of the same kind of diplomacy and laws that govern such relations in Europe of the nation-state. Structurally, the borders of Africa pose as much obstacle to international cooperation efforts as have their counterparts in Europe (Zartman 1965; Touval 1972).

African and Africanist scholars are quick to point out that the borders of modern African states are artificial; they were arbitrarily drawn with little or no regard for pre-existing territorial arrangements and have split up unified culture areas and natural zones or distinct ecosystems. From the viewpoint of the comparison in this chapter, the crucial point that is often missed in these familiar remarks is that in all such matters the borders of nation-states in Africa are really not different from the European ones.[3] It has been argued elsewhere, for example, that the phenomenon of artificially partitioned culture areas is as much a feature of borders and borderlands in Africa as they have been in Europe and the wider world of European-type nation-states (Asiwaju 1985).

In France, for example, the ethnic structure of the state, in spite of the centuries of centralist-state tradition, is not in any essential sense different from the structure of the independent states that have grown out of its former African colonies. With particular reference to the issue of partitioned culture areas astride the borders and the corresponding internal multi-ethnic situation of the state, Peter McPhee has most aptly observed:

> The territorial map of France does not correspond to an ethno-cultural map, either within the hexagon or along its borders. To the north-east the border slices into the Flemish-ethnic entity [the rest of which is to be found in the adjacent area of Belgium]. In the South, the border cuts through two Iberian lands: the Basque country (Euskadi) and Catalonia [each shared with Spain]. Corsicans speak Italian dialects. Lower Brittany has a million people using the Celtic language. There are at least ten or twelve million who know something of Occitan. Inversely, the territorial map of France does not include the French-speakers of Belgium and the Bernois Jura. Franco-Provence is spoken between St. Etienne and Fribourg, Grenoble and Lors-le-Sannier, even in the Val d'Aosta. (McPhee 1980)

The phenomenon of shared populations and related natural resources, important in any discussion of transboundary economic cooperation, was widespread in Europe – as can be very easily illustrated by the numerous instances on the borders of Switzerland, Italy, Yugoslavia, Turkey and so on. These European examples are replicated by the several parallel examples in Africa.

The features of a border region, such as we have tried to summarize, make it respond well to its typical and functional definition as an area of national and cultural interface, a region of social and economic interpenetration in which the functions and influence of the state on the one side of a given boundary normally fade gently into the sphere of its neighbour. This has to be in view of the fact of shared human and material resources and associated activities and interests spanning the border.

In spite of this fact, the policy vision of the situation in Europe up to the 1960s was more or less as it is today with regard to most states of Africa where borders and borderlands have continued to be viewed largely in conflict and tension terms. The notion of sovereignty (as expounded in the thoughts of eighteenth century European political philosophers) and the rather exclusive use of diplomacy and international law as sole instruments for dealing with international relations, including those affecting common borders and border regions, have continued to be emphasized at the expense of positive efforts at international cooperation. Political concerns, particularly those relating to state security, are viewed as far more important than issues of economic cooperation and overall social welfare. In Africa, as in Europe of the pre-World Wars era, most states are faced with threats of instability and this fact has come to exercise a commanding influence on the more or less permanent state of tension along most of the borders.

This chapter draws attention to the situation in Europe where the position has eased somewhat within the last two decades. This change has been due to a conscious and collaborative effort on the part of the nation-states. Rather than continue to view borders in conflict and tension terms, the new attitude in Western Europe has allowed for a new approach which now places emphasis on the more positive issues of transborder cooperation, planning and development. Borderlands have come to be viewed in terms of human needs and material developmental possibilities. Contrary to the disposition of the traditional diplomatic approach, by which the often distantly based national governments monopolized control, the new strategy allows such borderlands problems that are local to be so studied and resolved. It is realized, for example, that regional planning and development by any national government is impossible without adequate consultation with national authorities on the other side of the border. It is especially impossible for a local or regional community or authority in a border area to undertake coherent local development programmes without consultation with and possible input by counterparts on the other side of the border. What is said about development extends to issues of security, including the question of law enforcement, in the border areas.

In such a changing condition, the old ideology of national self-interest is being strongly persuaded to yield to an emergent alternative doctrine of mutual necessity and functional relationship between peoples and governments on both sides of a border. National sovereignty and associated international boundaries are not abandoned, but encouragement is given for borders to function less as lines of division and exclusion than as points of contact and mutual inclusion. In Europe, the cradle of nation-state and borders, transborder informal linkages which have been inspired by the fact of shared human and material resources straddling several of the borders, have been accorded formalized and internationally legalized status in recent years. Through processes and developments which have spanned the last twenty or so years, the member states of the Council of Europe, which embraces the whole of Western Europe, have between 1979 and 1982 endorsed the European Outline Convention on Transfrontier Cooperation between Territorial Communities or Authorities (Council of Europe 1982). The Convention is a major triumph of 'micro-diplomacy'; the implementation of its provisions would be the final demonstration of the incomparable advantage of international socioeconomic integration at the grassroots level.

The Outline Convention is indisputably the most important result of the series of experiments in international socioeconomic cooperation at the local level. The Convention and its antecedents issued out of a persistent search for solutions to regional and local problems. Success has depended at every stage on the support of the border communities themselves and, as Hansen has correctly observed, 'even though economic development is the stimulus to transboundary cooperation, such efforts can be facilitated by the presence of similar linguistic or ethnic groups on both sides of the border' (Hansen 1983). In this regard, it is pertinent to emphasize the leading role of the peoples along and across the Rhine valley where, as in Africa, 'borders represent more the results of past conquests and diplomatic arrangements than natural geographical barriers' (Hansen 1983).

The interest of the local people was so easily enlisted because in

Europe, as elsewhere, national governments generally prove negligent or negative in their attitude to the affairs of their border regions. National governments, often based at considerable distance from the borders, generally vacillate in dealing with problems affecting these sub-national areas because 'the only institutionalised means of communication across borders was international law and diplomacy' (Hansen 1983). This is often compounded by the usual bureaucratic red tape of state officials. Central governments, especially those of centralist states like France, impeded transborder cooperation 'because of fear that the sovereignty of the nation-state would be compromised'. The social, economic and political marginalization of border regions which generally result from this style of handling by national governments, often leave the border communities with the impression that their salvation can only lie in their own hands.

In the circumstances in which border communities in Europe found themselves, the need for self-reliance was too obvious to require being preached about from outside. In a situation of economic neglect, such as is so clearly reflected in the influential 'location theory' and 'growth pole' literature which commonly stress the disadvantage of border regions in matters relating to industrial location and international trade, and in situations where border communities must cope with problems of conflicting official mandates, resulting from manifestations in border regions of the differing political, legal, administrative, economic and cultural traditions and institutions of the nation-states in contact, the local people often have no choice but to rely on their own steam (Hansen 1972). Here lies the significance of the informal linkages which generally characterize border relations. In the absence of officially sanctioned mechanisms, sufficiently informed and adequately sympathetic towards the special requirements of the border regions, the informal network is the only mode left by which the local people can circumvent the problems of the border. These often include the barrier effects which generally stand in the way of border peoples wanting to take obvious 'advantage of potential complementarities in public services and facilities as well as potential scale and agglomeration economies' (Hansen 1972).

Not surprisingly, then, the experiments in transboundary cooperation which preceded the establishment of the European Outline Convention were all results of local initiatives. They include such organizations as the Regio Basiliensi created in 1963, the Euregio, launched in 1970, and the Conference of Upper Rhine Valley Planners, begun in 1979. The others are the federal-type Association of European

Border Regions, based in Bonn, western Germany, the Committees for the Promotion of Alpine Region Cooperation with headquarters in Turin, Italy and the Liaison Office of European Regional Organizations based in Strasbourg, France.

The Regio Basiliensi, one of the early examples of these transboundary cooperative efforts, was directly inspired by both the business and civic élites in Basel, Switzerland, who felt the need to do something positive about the increasingly pronounced economic marginalization of the city, due largely to its location at the north-west corner of Switzerland (Riner 1981; Schmid 1981; Regio 1981). But, as already hinted, the border situation made it impracticable for any coherent planning to be undertaken for Basel without due cognizance taken of the city's status as a trinational agglomeration with parts in the adjacent areas of North-western Switzerland, Upper Alsace in France and 'Sudbaden' in Germany. Indeed, for a more complete view of the planning needs of the area a good note was taken of the wider regional setting of the city. Regio Basiliensi then came to have as the area of its operation the entire historical 'Regio' which is defined as 'the European border area of the Upper Rhine between the Jura, the Black Forest and the Vosges ... [which] has more than two million inhabitants' (Riner 1981; Regio 1981).

In the words of Dr Hans Briner, the founder and Secretary General of Regio Basiliensi, 'its purpose is the planning and encouragement of economic, political and cultural development of this trinational area' (Riner 1981; Regio 1981). Funded through contributions initially made by private business and the concerned Swiss Cantons, Regio Basiliensi was founded on centuries of traditions of 'close cultural and economic ties', most of which antedated the birth of France, Germany and Switzerland as nations. Alemanic, a German dialect, is the common local language. The preservation of this common history and culture has been stated as an objective for the 'coordinated planning and harmonization in the individual regions' (Riner 1981; Regio 1981).

Regio Basiliensi's achievements are quite impressive. Quite apart from its series of surveys and recommendations on coordinated transportation and its environmental impact studies with particular reference to the implications and consequences of the nuclear plants located in the area, Regio Basiliensi has provided the brain behind the creation of the Basel–Mulhouse airport, built with Swiss money but located on French soil and jointly managed by both countries. Of crucial importance are studies of the opinions of the local people as the proper basis for the transboundary cooperation efforts in the trinational area.

The Euregio founded in 1970, constituted another early example of transboundary cooperation in Europe (Hansen 1983). Essentially a federation of three pre-existing associations of municipalities in the Germany–Netherlands border region, the Euregio operates through a joint international secretariat and a parliamentary council. The main objectives are 'to develop common policies and programmes in socio-cultural and economic matters and to promote the principle of a genuine European transborder region, priority being given to the development of a single transborder region, not two border sub-areas' (Council of Europe 1980).

The Conference of Upper Rhine Valley Planners, which was launched in 1979, represented a major endeavour to promote large-scale transboundary planning on an informal level. While the focus of the Conference has been on issues affecting West Germany, France and Switzerland, a fact that would have made it a mere duplication of the Regio Basiliensi, the Conference's interest covers the whole area of the Rhineland from Basel to Frankfurt. Again, like the Regio Basiliensi, the Conferences's main impact has been in terms of the detailed analyses it has sponsored on such shared environmental issues as water and air pollution and undesirable urban pattern. There are also studies of planning in several matters including land use, economic and demographic trends, and transportation.

The Association of European Border Regions, based in Bonn, advanced the trend of federation already noted in the discussion of both the Euregio and the Conference of Upper Rhine Valley Planners. It consists of fifteen border region organizations most of which belong to the Rhine valley. Of similar structure and function is the Committee for the Promotion of Alpine Region Cooperation based in Turin, Italy. Both the Association of European Border Regions and the Committee for the Promotion of Alpine Region Cooperation are closely linked to the Council of Europe based in Strasbourg, France.

The European Outline Convention should be considered as the logical conclusion of this whole business of upward movement for transboundary cooperation in Europe. It is by far the most comprehensive and all embracing. For those of us in Africa, the Convention represented a noteworthy example of how to build international socio-economic cooperation at the level of a whole continent or any of its major sub-regions upon solid local foundations. The Council of Europe, which produced the Convention, was the first and, with twenty-one member states, it is still the largest Western European political organization to be formed since the traumatic experience of the Second World War.

Founded in 1949, the Council operates through a parliamentary assembly and the Committee of Ministers which function as its deliberative and executive organs respectively. The Committee of Ministers decides on recommendations submitted to it either by the Assembly or the committees of government experts that may be set up from time to time. The decisions take the form of recommendations to governments or formal conventions or agreements. Member states who ratify these decisions are legally bound by them.

Given the intense and widespread nature of transborder cooperative movements and organizations in Europe, it is not a surprise that the subject has been a matter of increasing interest to the Council. Accordingly, a number of organizations created within its framework with assignments specifically related to the question of international cooperation in border regions. The two most notable of such Council of Europe organizations are the European Conference of Ministers responsible for Regional Planning and the Conference of Local and Regional Authorities of Europe (CLRAE), the latter being the only such body in Europe which represents local land regional authorities in their relations with international institutions.

Because of its intimate link with, and direct knowledge of the aspiration of local communities, including those in border areas, the CLRAE has proved the most vital organ of the Council of Europe and one which has done more than any other body to give rise to the Outline Convention. It did not just encourage central governments to recognize the need for transborder cooperation; more importantly, 'it has emphasized that the organization of such cooperation required the close participation of elected representatives of border region and municipalities'. CLRAE's direct involvement with border regions has left it without any doubt that such regions and authorities are the loci where problems of international cooperation are most keenly felt. CLRAE is, therefore, the organ which has made clear to the Council of Europe that regional planning and development in a border region is completely unrealistic without a systematic consultation between planning authorities on both sides of the boundary. The Committee's view of transboundary relations as a process of five progressively complex phases of information exchange, mutual consultation, active collaboration, harmonization of planning and integration of planning has had a strong impact on the notion of progression manifest in the Outline Convention and its Appendix.

The Outline Convention was passed by the Council of Europe's Parliamentary Assembly in 1979 and on 21 May, 1980 in Madrid it was opened for signature. It has now been signed by Austria, Belgium, Denmark, France, Ireland, Italy, Luxembourg, The Netherlands, Norway, Sweden, Switzerland and Germany. There are, altogether, twelve articles plus an appendix containing five draft 'Model Inter-State Agreements' and six model 'Outline Agreements, Statutes and Contracts between Local Authorities'.

The provisions of the Convention oblige each signatory member state 'to facilitate and foster transfrontier cooperation between territorial communities or authorities within the jurisdiction of other Contracting Parties' (Article 2). The purpose of the Convention was to achieve 'concerted action designed to reinforce and foster neighbourly relations between territorial communities or authorities within the jurisdiction of two or more Contracting Parties' (Article 2:1). Although each signatory member state retains as always its sovereign rights on matters of international relations, there is an undertaking by each 'to resolve any legal, administrative or technical difficulties liable to hamper the development and smooth running of transfrontier cooperation' (Article 4).

The draft model agreements between states or between local authorities are to assist such states and authorities to conclude supplementary agreements that will facilitate the execution of the provisions of the Convention. Sixteen broad areas of transborder cooperation have been identified on the basis of existing experience. These include urban and regional development; transport and communications; energy; nature conservation; water conservation; education, training and research; public health; culture; leisure and sports; mutual assistance in disaster; tourism; commuter labour; economic projects; improvement in agriculture; social facilities; and miscellaneous issues (Article 6 of the second Model Inter-State Agreement on Transfrontier Regional Consultation).

The European experience in transboundary planning, as outlined above, may be considered relatively too brief to become an authoritative reference point. In fact, the Outline Convention was not ratified by France, understandably the most reluctant and the last Council-of-Europe member to do so, until 1984. However, the history and substantial achievements of the Convention as a treaty with a firm rooting and foundation in the geographical realities on the ground and the resultant spontaneity of support by the signatory powers must be viewed as favourable and sufficiently reliable indicators of success (Alois 1984). In any case, the urgent need in Africa for an alternative strategy which would make current efforts at regional economic integration more relevant and more demonstrably meaningful to the predominantly rural and local peoples of the continent (so well represented by the border communities) oblige us in Africa to take serious note of the European experience.

It is already worthy of the attention of Africa that in spite of traditional attachment to the notion of sovereignty and the use of conventional methods of diplomacy and international relations, including international economic cooperation, the governments of Western European countries, most of them members of the Council of Europe, have mobilized their concerted effort for a new kind of diplomacy that allows local and regional authorities on the borders of the nation-states to function as agents of relations and cooperation between them. The era of borders that divide and separate thus appears to be giving way to a new era of borders that join (Haddox 1982; Brown and Shue 1983). As the informal linkages across European international boundaries are accorded formalized status, the discrepancy and gaps between the boundary-maintenance policy of nation-states above and the boundary-disregarding attitudes and behaviours on the ground are being reduced, if not eliminated. There is in this new development a demonstrable effect which gives us in Africa an opportunity not only for a comparable adjustment of policy to the realities of relations on our borders but also a new chance to recover what was lost at the time of the establishment of the frictioned boundaries, viz., the pre-colonial concept of boundaries as zones of mutual contact (Asiwaju 1983).

We have already pointed to the structural and functional similarities between borders and border regions of Europe and those of Africa. There is, therefore, no doubting the fact that the human and material resources for achieving transboundary cooperation in Africa are no less than those in Europe. The border regions of Africa may and do vary not only from those of Europe but from one another in terms of local details. Nevertheless, the basic prospects and problems are substantially the same. With particular reference to the question of transboundary cooperation, Africa has all, and perhaps even more than was required in Europe.

First is the fact of abundant human resources. This point can be simply illustrated by the widespread presence of peoples and communities of the same culture on both sides of every border. My recently completed research on this subject provides some conclusive evidence: a detailed checklist of African border culture areas shows that each of the one hundred and three international boundaries on the continent features this type of 'population overhead'.[4] The proverbial neglect of border areas, a pronounced feature of the African situation characterized by a general absence of industrial activities and urbanization processes, has aided the accentuation of the primordial cultural pull and intra-group relations across the borders (Asiwaju 1983; 1985). The sense of rejection and disregard for the borders in partitioned cultural areas of Africa is probably greater than in the industrialized and relatively individualistic societies of Europe. The peoples of Africa, those at the border not exempted, are known for their mistrust, expressed in many instances in the form of armed resistance, of European colonial establishments, including the establishment of the colonial boundaries. Against this background, it is easy to see how the cultural areas astride such boundaries have been conducted to domesticate, if not totally neutralize, the locally judged 'harmful' effects of the alien cultures and the differing state mandates which criss-cross the borders.

The ethnic ties across African borders are often reinforced by the interlocking character of the border settlement pattern. Here again, the details vary between the two border situations under discussion but the substance of the matter is once more the same. Thus, whereas urbanization along borders is a major issue in transborder relations in Europe, there is perhaps no location outside the singular case of Brazzaville (Congo) and Kinshasa (formerly Leopoldville, in Zaire) where settlements of standard urban sizes are found in a twin-type location along any African binational boundary. However, there are equivalent medium-sized urban centres not directly on the boundaries but within the confines of the borderlands, as well as innumerable village-level communities of the same type of dual location along several African borders.[5] As in Europe, there is the same degree of interdependence in social and economic matters and a comparable degree of informal linkages between the communities across the borders. In a number of instances, as Dr Mills has discoved in his study of the Nigeria–Benin case, the frontier village communities are experiencing a remarkably high rate of growth, enough to justify the advance attention of experts concerned with border environmental questions (Mills 1973).

African border communities and local authorities face the same kind of social, economic and political problems that stimulated development from the informal to formalized transborder relations in Europe. Quite apart from questions relating to shared land and related natural resources, which call for transborder planning, experiences of governments in border areas point to the need for joint action. Examples abound of government rural development programmes on one side of the border, falling short or becoming inadequate precisely because the services, provided for the nationals on the one side of the border could not be shut against their kinsmen from the other side. A situation like this has, for example, generated complaints by the Zambian government about Mozambican and Malawian Chewa and Ngoni kinsmen who overburden the medical, agricultural and educational facilities provided under the state's rural integrated scheme for kinship groups resident on the Zambian sides of the boundaries with these neighbouring states (Phiri 1980).

A description of the policy potentials in Africa would be incomplete if it did not include some statements as to whether or not there are in Africa institutional infrastructures or frameworks with the level of capacity which has enabled the Council of Europe to play its role in the European case. There is, of course, no doubt that there are numerous intergovernmental organizations in the continent, but for the purpose of this discussion, there appears to be no better choice than the Organization of African Unity (OAU). The human and material resources necessary for the adoption of a transboundary planning and coordination strategy have been found to be a truly African-wide phenomenon. What is required is an organization with political leverage at the level of the entire continent itself. Aside from the current anxiety about the OAU's commitment to its own preservation and survival, there is no doubt about its capacity to play the roles analogous to those of the Council of Europe. If adequately informed and advised, the OAU can resolve to aid the promotion of adjoining borders in the same way as it has resolved to maintain the boundaries as they were at the independence of member states. The same concern for continental unity, peace and overall development which led to the resolution for the status quo in 1963 is today all the more compelling in relation to the decision to devalue the barrier effects of the boundaries. African equivalents of the European Conference of Ministers responsible for Regional Planning and especially the CLRAE can be easily added to the existing organizations with the OAU.

Given the fact of the immense possibilities for transboundary planning in Africa, the question then arises as to what should be done to translate the potentials into actual policy. Several considerations easily come to mind, but perhaps the most important and urgent relates to the need for a massive and systematic public enlightenment programme. This is necessary to bring about a desirable change in the attitude and outlook of the political and bureaucratic élites at all levels, including those of the border regions themselves. Special emphasis is placed on the local political and bureaucratic élite if only to call attention to the

71

extent to which Africa has lagged behind trends in Western Europe and some other parts of the world, notably North America, in spite of having comparable resources. Whereas in Western Europe and say, the USA–Mexico borderlands, informal linkages across borders have for long involved local authorities on both sides of borders, in Africa cross-border informal linkages have remained for the most part an affair of the members of the border communities and their traditional rulers without the involvement of their local governments or local administrations manned by the western-educated élite (D'Antonio and Form 1965; Sloan and West 1976; Regio 1981; Asiwaju 1984a). This point is especially manifest in situations such as prevail along the four landward borders of Nigeria where, as elsewhere in West Africa, an officially English-speaking federalist state is in daily contact with French-speaking centralist neighbours, and 'local governments' in the one state are in direct juxtaposition with the 'local administration' of its adjacent neighbours.[6]

To ensure success of the public enlightenment programme, the need for a new approach to borderlands research is more than obvious. This calls for a specially collaborative effort. Routinely, the recognition of borderlands as distinct regions compels the use of a multidisciplinary approach to their study. This point is all the more pressing since the view of development and cooperation taken in this chapter is one of complete integration, not limited to trade and market. Encouragement should therefore be given for teamwork involving experts of appropriate research orientation and interests in the related disciplines of the humanities, social sciences, law, environmental design (with particular reference to environmental impact studies of border situations), education, public health, agriculture and the natural sciences (Asiwaju and Adeniyi 1989). The second level of collaboration relates to institutions focusing on the various borders or distinct segments of particular borders and border regions. Such institutional collaboration must connect relevant research centres both within individual nation-states and across their borders (Asiwaju 1984b). This binational dimension is especially demanded by the need adequately to inform policies on bi-national relations, good neighbourliness and international cooperation.

NOTES

1 'Border', 'boundaries' and 'frontiers' are contested concepts, but for the purpose of this chapter, the three will be taken as synonyms and used to refer to the *line* of demarcation between any two sovereign states. 'Border regions',

'borderlands' and 'frontier zones' will be the reference to the sub-national areas on both sides of the binational line of demarcation.

2 Based on the model of Switzerland as a typical example, all member-states of ECOWAS have predominantly 'border' characteristics. These include not just small-sized states like Benin, Togo and The Gambia or the land-locked Sahelian states of Niger, Upper Volta (Burkino Faso) and Mali. There is also Nigeria, the biggest of the states in the sub-region, with fifteen of its present twenty-one constituent states directly linked with the international boundaries.

3 It is, for example, a matter of great fascination to compare the responses of the Catalans, 'an ethnic group, neither French nor Spanish' to the parallel socialization processes by France and Spain in the Cerdanya Valley arbitrarily divided into two by the Franco-Spanish border in the Eastern Pyrenees with those of the Yoruba, also an ethnic group neither French nor English, arbitrarily divided into two by the Anglo-French Nigeria–Dahomey colonial boundary in Western Yorubaland (Asiwaju 1977; Sahlins 1989).

4 Examples of 'partitioned Africans' include the Yoruba and the Borgu astride the Nigeria–Benin boundary, the Hausa across Nigeria/Niger, the Ewe of Ghano/Togo, the Wolof of Senegambia in West Africa; the Somali of Somalia/Ethiopia/Kenya/Djibouti in North-west Africa; the Massai of Tanzania/Kenya, the Alur of Uganda/Zaire, the Kongo of Zaire/Congo/ Angola, the Cokwe and the Lunda of Angola/Zaire in Central and East Africa; the Chewa-Ngoni of Zambia/Malawi/Mozambique, the Name and Ba-Tewana of Angola/South Africa and Botswana/South Africa respectively – all of Southern Africa. Each of these culture areas is sufficiently large to constitute units for transboundary regional planning.

5 Examples of border settlements of interlocking location include such prominent Nigerian cases as Badagry and Porto Novo, Imeko and Ketu, Bussa and Nikki along the Nigeria–Benin border; the two Daura, and Katsina and Maradi on the Nigeria–Niger boundary and other dual villages on the Nigeria–Cameroon border.

6 The concept of 'local government' denotes a degree of local autonomy that does not form part of the concept of 'local administration'; it is, therefore, usual to use the first in reference to local or regional authorities in federal states and the latter for equivalent units in centralist states.

REFERENCES

Alois (1984) 'The international development of transfrontier cooperation: achievements and prospects: implementation of the European Outline Convention and Authorities', 3rd European Conference on Frontier Regions, Barken, Sept.

Anene, J.C. (1970) *The International Boundaries of Nigeria, 1885–1960*, London: Longman, Ibadan History Series.

Asiwaju, A.I. (1977) *Western Yorubaland under European rule, 1289–1900: A Comparative Analysis of French and British Colonialism*, New Jersey: Humanities Press.

—— (1983) 'The concept of frontier in the setting of state in pre-colonial

73

Africa', *Presence Africaine*, 127/128.

—— (1984a) 'Informal linkages across borders: the African experience' paper presented at the International Seminar on 'Problem-solving along borders: a comparative perspective', Centre for InterAmerican and Border Studies of UTEP, March.

—— (1984b) *Artificial Boundaries*, New York: Civiletis International.

—— (ed.) (1985) *Partitioned Africans: Ethnic Relations across Africa's International Boundaries, 1884–1984*, London: Hurst.

—— (1986) 'Borderlands as regions: lessons of the European transboundary planning experience for international economic integration efforts in Africa', in Owosekun, A.A. (ed.) *Towards an African Economic Community*, Ibadan: NISER.

Asiwaju, A.I. and Adeniyi, P.O. (eds) (1989) *Borderlands in Africa: A Multidisciplinary and Comparative Focus on Nigeria and West Africa*, Lagos: University of Lagos Press.

Brown, P. and Shue, H. (eds) (1983) *The Border that Joins*, New Jersey: Rowman and Littlefield.

Brownlie, I. (1979) *African Boundaries: A Legal and Diplomatic Encyclopaedia*, London: Hurst.

Council of Europe (1980) *Conference of Local and Regional Authorities of Europe: Report on Transfrontier Cooperation in Europe*, CPL 15: 6, Strasbourg, cited in Hansen (1983).

Council of Europe (1982) *European Outline Convention on Transfrontier Cooperation between Territorial Communities or Authorities*, Strasbourg, European Treaty Series: no. 106.

D'Antonio, W. and Form, W.H. (1965) *Influentials in Two Border Cities*, University of Notre Dame Press.

Duchacek, I.D. (1986a) 'International competence of sub-national governments: borderlands and beyond' in O.J. Martinez (ed.) *Across Boundaries: Transborder Interaction in Comparative Perspective*, El Paso: Western Texas Press.

—— (1986b) *The Territorial Dimension of Politics Within, Among and Across Nations*, Colorado: Westview Press.

Duchacek, I.D., Latouche, D. and Stevenson, G. (eds) (1988) *Perforated Sovereignties and International Relations: Trans-Sovereign Contacts of Subnational Governments*, Connecticut: Greenwood Press.

Haddox, J.H. (1982) 'The border: a place to live, a place to learn', Faculty Research Award Lecture, University of Texas at El Paso.

Hansen, N. (ed.) (1972) *Growth Centres in Regional Economic Development*, New York: Free Press.

—— (1981) *The Border Economy: Regional Development in the Southwest*, Austin: University of Texas Press.

—— (1983) 'European transboundary cooperation and its relevance to the United States–Mexico border', *Journal of the American Institute of Planners* 49, 3: 336–43.

MacEwan, A.C. (1971) *International Boundaries of East Africa*, Oxford: Oxford University Press.

McPhee, P. (1980) 'A case study of international colonization: the Francization of Northern Catalonia', *Review* III, 3: 400–1.

Mills, L.R. (1973) 'The development of a frontier zone and border landscape along the Dahomey–Nigeria boundary', *The Journal of Tropical Geography* 36: 44.

Owosekun, A.A. (ed.) (1986) *Towards an African Economic Community*, Ibadan: NISER.

Phiri, S.H. (1980) 'Some aspects of spatial interaction and research to government policies in a border area: a study in the historical and political geography of rural development in the Zambia/Malawi and Zambia/Mozambique frontier zone (1970–79)'. Unpublished Ph.D. thesis, University of Liverpool.

Quitin, J.M. (1973) *European Co-operation in Frontier Regions*, Strasbourg: Council of Europe.

[Regio] Service de Coordination Internationale de la Regio (1981) *General Information*, Basel: Service de Coordination Internationale de la Regio.

Riner, H.J. (1981) *Coordination of Regional Planning Across National Frontiers: Regio Test Case – Switzerland/France/Germany*, Basel: Service de Coordination Internationale de la Regio.

Sahlins, P. (1989) *Boundaries: The Making of France and Spain in the Pyrenees*, Berkeley: University of California Press.

Schmid, H.P. (1981) *The Transfrontier – Impact of Nuclear Power Stations in Basle-Region*, Basel: Service de Coordination Internationale de la Regio.

Sloan, J.W. and West, J.P. (1976) 'Community integration and policies among élites in two border cities: Los Laredos', *Journal of InterAmerican Studies and World Affairs* 18: 451–74.

—— (1977) 'The role of informal policy making in United States–Mexico border cities', *Social Science Quarterly* 58, 2: 277–82.

Touval, S. (1972) *The Boundary Politics of Independent Africa*, Harvard: Harvard University Press.

UN–ECA (1984) *Strengthening Economic Integration in West Africa*, Addis Ababa: ECA.

Zartman, W.I. (1965) 'The politics of boundaries in North and West Africa', *Journal of Modern African Studies*, 3 August.

7

ANTARCTICA
The Antarctic Treaty System
after thirty years

Peter J. Beck

INTRODUCTION

In June 1991 the Antarctic Treaty System (ATS) celebrated thirty years of operation. This anniversary conveyed the impression of durability and permitted Antarctic Treaty Parties (ATPs) to reaffirm their continuing support for a 'successful' regime:

> The Antarctic Treaty has for thirty years united countries active in Antarctica in a uniquely successful agreement for the peaceful use of a continent.... The Antarctic Treaty provides an example to the world of how nations can successfully work together to preserve a major part of this planet, for the benefit of all mankind, as a zone of peace, where the environment is protected and science is pre-eminent.... The strength of the Antarctic Treaty continues to grow and parties to the Treaty now represent over 70% of the world's population. (Declaration by ATPs, October 1991)[1]

The parties to the Treaty comprise a diverse and formidable range of countries possessing differing political, economic and social complexions, including developed and developing, aligned and non-aligned countries, the major nuclear powers, and all permanent members of the Security Council (Table 7.1). Contemporary debates about Antarctica have polarized in part around the merits of the ATS as a management mechanism. ATPs remain committed to the ATS, whereas the critics, led by Malaysia, advocate the need for a replacement regime reflecting 'the true feelings of the Members of the United Nations or *their just claims*' (UNGA A37/PV10 Mahathir 29 September 1982: 18–20). The 1992 UN session saw no progress towards a consensus view.

Table 7.1 The growth in membership of the Antarctic Treaty System

Consultative Parties (ATCPs) – original signatories and states adjudged to perform 'substantial research activity' in Antarctica entitled to a decision-making role at Antarctic Treaty Consultative Meetings.

1959–61[1]	Argentina			
	Australia			
	Chile	Claimant		
	France	States (7)		
	New Zealand		Original	
	Norway		Treaty	
	United Kingdom		Consultative	
			Parties (12)	
	Belgium			
	Japan			
	South Africa			
	USA			
	USSR			Consultative
				Parties (26)
1961	Poland		1977	
1974/79	Germany (1990)[2]		1981/87	
1975	Brazil		1983	
1983	India	Non-	1983	
1983	China	Claimant	1983	
1980	Uruguay	ATCPs	1985	
1981	Italy	(19)	1985	Additional
1982	Spain		1987	Consultative
1984	Sweden		1988	Parties (14)
1984	Finland		1988	
1981	Peru		1989	
1986	Republic of Korea		1989	
1967	Netherlands		1990	
1987	Ecuador		1990	

Non-Consultative Parties – recognize the validity of the Antarctic Treaty and have observer status at meetings. It is often a stepping-stone towards ATCP status.

1962	Czechoslovakia	
1965	Denmark	
1971	Romania	
1978	Bulgaria	
1981	Papua New Guinea	
1984	Hungary	Non
1984	Cuba	Consultative
1987	Greece	Parties (15)
1987	Democratic People's Republic of Korea	
1987	Austria	

Table 7.1 continued

1988	Canada	
1989	Colombia	
1990	Switzerland	
1991	Guatamala	
1992	Ukraine	

Total Number of
Parties: 41

Notes: ¹The original twelve ATCPs signed and ratified the Treaty between 1959 and
1961. Other dates in the left-hand column indicate the date of accession to the
Treaty, while those in the middle record admission to ATCP status.
²In 1990 the unification of Germany involved the merger of two ATCPs – the
Federal Republic of Germany (1979: ATCP 1981) and the German Democratic
Republic (1974: ATCP 1987).

THE REAL IMPORTANCE OF THE UNIMPORTANT

One area for controversy concerns the ownership of Antarctica. During
recent centuries the political framework of the world has been organized
upon the basis of nation-states. This trend, though challenged by
integrative tendencies, shows few signs of abating. The world has been
parcelled out among states and delineated by clearly demarcated
frontiers. Sovereignty – the right to exercise jurisdiction over a territory
to the exclusion of other states – has been claimed over virtually every
part of the world's land surface. More recently, the UN Convention on
the Law of the Sea (UNCLOS), which was signed in 1982 but has yet to
come into effect, clarified maritime jurisdiction concerning 200-mile
Exclusive Economic Zones, while declaring the deep seabed as the
common heritage of mankind.[2]

According to Professor Alan James (Keele University):

It is possible to say that, jurisdictionally speaking, there is never
any doubt about where one stands, and that one always stands on
the domain of a single sovereign state. The exceptions are so small
or so literally out of the way as to prove this rule. (James 1984:
16–17)

The student of international boundaries, aware of numerous border
disputes, might disagree, while pointing to the way in which serious
conflicts have arisen out of sovereignty problems, as illustrated recently
by the 1991 Gulf War centred on the Kuwait–Iraq issue. The year 1992
marked the 10th anniversary of the 1982 Falklands War, which was

78

caused by the long-standing Anglo-Argentine rivalry over the Falkland Islands (Beck 1988a). The continuing impasse over this archipelago highlights the fact that several territories, particularly in Latin America, remain subject to rival claimants. The history of this dispute displays also the capacity of disagreements about so-called 'dots on the map' to have serious consequences, even if most disputants manage either to co-exist peacefully and/or to move towards a settlement (Beck 1991a).

THE ANTARCTIC DIMENSION: A POLE APART?

The Falklands War also prompted an enhanced focus upon another often ignored region, the nearby continent of Antarctica. An appreciation of Anglo-Argentine differences regarding the legal status of a portion of this vast southern continent led some politicians and commentators to articulate the possibility of an Antarctic dimension to the Falklands War (Beck 1986: 83–5).

Alan James suggests that 'the allocation of the world's land surface to sovereign states has reached virtually the ultimate point', excepting 'small' or 'out of the way' areas like Antarctica.

> There is no agreed division of the Antarctic continent, and the various competing claims to parts of it have, in effect, been put on one side under a treaty of 1959. (James 1984: 16–17)

This reference typifies the tendency of most international relations and geography specialists to interpret Antarctica as an exotic, uninhabited and 'out of the way' region isolated from the mainstream of international affairs. It is perceived in terms of images associated with polar explorers, like Scott, Mawson and Shackleton, that is, in a manner reinforcing the popular vision of an empty, inhospitable and obscure white continent to be viewed from afar. Thus, Antarctica is frequently presented as 'a pole apart' because of its peripheral location, geographical isolation, tardy discovery (1819–21), unknown nature, and pristine features (Quigg 1983; Sugden 1982). It is a cold, windswept and ice-covered continent, where climatic conditions tend to extremes, as evidenced by minimal precipitation in the interior, a record low temperature of −89.6°C, and high velocity winds. There are no indigenous inhabitants, only transient researchers and other visitors; indeed, man is essentially alien to a region characterized by a combination of blizzards, low temperatures and seasonal light regime. The long-standing tendency of governments and academics to treat Antarctica as a marginal region has been encouraged by cartography.

World maps tend either to show merely a small coastal section or to omit the continent altogether. A map centred on the South Pole establishes the vast size of Antarctica, whose area of 14 million sq. km exceeds the combined extent of either China and India or the USA and Mexico (Figure 7.1). It accounts for 10 per cent and 30 per cent of the land surface of the world and southern hemisphere respectively.

To some extent, the 1980s witnessed a change of attitude. Antarctica, though never becoming a mainstream issue, moved towards centre stage. It could no longer be totally ignored (Parsons 1987: 3). This transformation was encouraged by a growing recognition of the region's scientific importance and integral role in global environmental systems, speculation about its living marine and mineral resource potential, anxieties about the growth of tourism, and controversy regarding modes of management. There was also the 'Falklands factor',

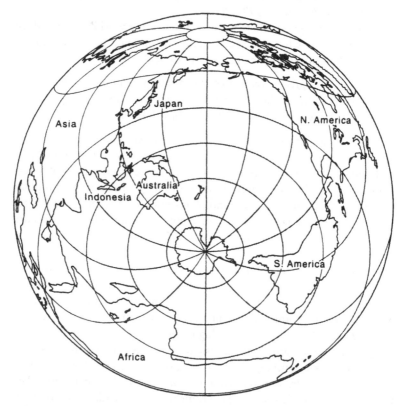

Figure 7.1 A cartographic expression of Antarctica

as mentioned earlier. Antarctica was deemed worthy of consideration by the UN as well as by a range of other international organizations, like the Non-Aligned Movement, the Organization of African Unity, the European Parliament, and the South Pacific Forum. Reports emanating from the UN, European Parliament and the Brundtland Commission, among others, contributed to the impression of a 'continent surrounded by advice' (Beck 1988b). Traditionally, Antarctica, whose affairs have normally been the preserve of diplomats and scientists, has not been an issue much visited by leading politicians. However, in recent years, the continent's enhanced visibility and place on the environmental political agenda explains the close attention of world leaders, like former President George Bush (USA), former President Mikhail Gorbachev (Soviet Union), Michel Rocard (French prime minister 1988–91), and Bob Hawke (Australian prime minister 1983–91) (Novost Press Agency 20 January 1990). In 1989 Margaret Thatcher (British prime minister 1979–90) employed a major environmental statement delivered to the UN to press the importance of Antarctica, which proved a key issue raised during her London discussions with Hawke during June 1989 (*The Times* 22 June 1989; *The Independent* 22 June 1989; *Daily Telegraph* 9 November 1989).[3]

These political developments were reinforced by the way in which non-governmental organizations (NGOs), including conservation groups such as the Antarctic and Southern Ocean Coalition (ASOC), Greenpeace International, the World Wide Fund for Nature, and the Cousteau Society, focused attention upon Antarctica as a major global problem testing the international community's environmental credentials and serving as a 'symbol of our collective ability to discipline ourselves' (Cousteau 1990; *Eco* 1990). After 1987, Greenpeace's 'World Park' Antarctic base offered novel opportunities for the NGOs' campaign to declare the last great wilderness on earth a 'world park' in which mining would be prohibited and other human activities (e.g. fishing, tourism and science) subjected to strict controls (*The Times* 9 January 1989; Greenpeace Press Release 9 January 1989; ASOC Press Release 19 November 1990).

A TREASURE ISLAND OR WASTELAND?

Antarctica's current economic value is low. Its economic future proves uncertain. Scientific data represent its prime export, and seem likely to remain so. As a result, it is easy to dismiss Antarctica as a wasteland unworthy of serious attention, despite a media tendency to advance a picturesque 'gold rush'/treasure island vision of the area (*The Sunday Times* 13 February 1983). This speculation was fuelled by the US Geological Survey's *statistical estimates* (1974) regarding a potential yield from Antarctica of 45 billion barrels of oil and 115 trillion cubic feet of natural gas as well as by a continuing focus upon the probable presence of strategic minerals like platinum (Wright and Williams 1974; Maarten de Wit 1985). Margaret Thatcher, looking back in 1989 to her government's enhancement of British research expenditure in Antarctica, stated that:

> I have always been interested in Antarctica ... Those fantastic, remarkable icy lands ... are not wastelands ... I did think that it was very very (*sic*) important for Britain ... We wanted to know more about the seas there, the wild life, and the mineral deposits.
> (Thatcher, BBC TV, *Nature Programme*, 2 March 1989)

No commercially exploitable deposits are known, even if it is reasonable to assume that minerals exist in and around Antarctica in the light of contemporary acceptance of the Gondwanaland hypothesis, according to which the region was formerly linked to geological strata in other continents. An accurate resource assessment is handicapped by a critical lack of knowledge, for geological research has failed hitherto to justify optimistic notions of tremendous riches. In any case, constraints imposed by ice, climate, distance from markets, technological in-adequacies, and conservation requirements mean that no mineral deposits likely to be of economic value in the foreseeable future, are known in Antarctica (Willan, MacDonald and Drewry 1990: 25–52). The extent of contemporary commercial interest is difficult to quantify, but realistically Antarctic minerals are resources of last resort (Larminie 1987: 176–81; Willan, MacDonald and Drewry 1990: 29). The recent adoption of the Protocol on Environmental Protection (PREP), which includes a permanent mining ban subject to review after fifty years (Articles 7 and 25), reinforces this view.[4]

By contrast, Antarctica's living marine resources already yield a financial return. In fact, the Southern Ocean has a long history of commercial whaling and sealing, as demonstrated by the manner in which a high proportion of the world's whale oil was derived from the seas surrounding Antarctica during the inter-war period. Relatively few whales are caught today – NGOs campaign for a complete ban – while in recent years annual catches of fish, squid and krill have proved relatively static at *circa* 0.5 million tons because of species vulnerability to over-fishing, the limited market for krill products, and the regulatory

measures imposed by the Convention for the Conservation of Antarctic
Marine Living Resources (CCAMLR) regime.[5] Nevertheless, commentators have often speculated about potential catch figures varying between
four million and 150 million tons per annum (Shackleton 1982: 65–82;
Beck 1986: 214–18).

Tourism, one of the world's fastest expanding industries, represents
the most dynamic economic growth area. The precise potential of
tourism is difficult to estimate, but already Antarctica's unique wilderness and scenic qualities are attracting an ever-increasing number of
tourists (2,581–6,500 per annum: 1987–92) and non-governmental
expeditions searching for new experiences and adventures (Enzenbacher
1992; Beck 1990a: 343–5). Numbers are expected to increase further,
as implied by recent proposals for the construction of airstrips and
hotels; in fact, some predict an eventual annual total of 25,000
(Australian House of Representatives Standing Committee on Environment, Recreation and the Arts 1990: 4–5; DASETT 1988: 17).

The 1990s seem likely to witness an escalating, albeit relatively low,
level of human activity in and around Antarctica in the spheres of
science, fishing and tourism. It must be questionable to continue
treating Antarctica as a 'wasteland'. In the meantime, economists are
taking a closer interest in the region, and tourism is beginning to
concentrate minds about possible conflicts of interest, such as between
conservation and development or prohibition and regulation (Beck
1991b: 34–5).[6] The basic problem has been articulated by Edmundo
Vargas, the Chilean foreign minister:

> We want a clean Antarctica, but we also want an Antarctica that
> is useful to man. That is why we are faced with the challenge of
> reconciling an Antarctica free from pollution with one open to
> human activity. (quoted in Beck 1991b: 37–8)

THE LATE 1950s AS A TURNING-POINT

The late 1950s proved a major turning-point for the southern polar
region. This period saw the major scientific programme conducted in
Antarctica as part of the International Geophysical Year (IGY 1957–8),
the creation (1958) of the Scientific Committee on Antarctic Research
(SCAR, known initially as the Special Committee on Antarctic
Research), and the conclusion of the Antarctic Treaty (December 1959).

During the succeeding three decades the relatively successful
operation of the Antarctic Treaty System (ATS) rendered it easy to
forget the pre-1959 scenario, that is, a period characterized by

conflicting sovereignty claims, the serious over-exploitation of whales and seals, and obstacles to international scientific collaboration. Turning back the pages of history, there was indeed a time between the 1920s and 1950s when there existed – to quote Christopher Beeby, a New Zealand diplomat with long experience of Antarctic affairs – 'a massive dispute about sovereignty in Antarctica' possessing 'ample potential for tensions, disputes and, in the worst case, armed conflict' (Beeby 1991: 3–6). During the 1940s and 1950s overlapping claims explained the unstable relationship between Argentina, Britain and Chile punctuated by frequent diplomatic difficulties, fears of a naval confrontation, and occasional incidents (Figure 7.2). In February 1952

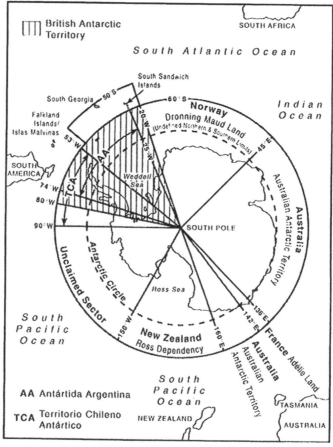

Figure 7.2 Territorial claims in Antarctica

Argentine military personnel – they were described in British official circles as 'trigger-happy South Americans' – fired shots to prevent the re-construction of a British base station at Hope Bay (Beck 1987: 16–17). This clash, though resolved through an exchange of notes, typified the conflict potential of the sovereignty issue (see Chapter 8 in this volume).

THE ANTARCTIC TREATY SYSTEM

The central feature of the politics, law, science and environment of Antarctica is the 1959 Antarctic Treaty. It offers a framework for the management of the region through a multilateral cooperative approach designed to ensure Antarctica's status as a zone of peace, a continent for science, and a special conservation area. The Treaty became effective in June 1961.

Henceforth, Antarctica was treated as a zone of peace defined by provisions for peaceful use, demilitarization, denuclearization, the shelving of sovereignty disputes, and the establishment of an inspection system (Beck 1990b). The significance of this aspect derived from the fact that the Antarctic Treaty, the first post-1945 agreement for the complete demilitarization of a sizeable geographical region, proved the forerunner for other nuclear weapon free zone agreements. In addition, the Treaty, employing an ingenious formula based upon a freezing of legal positions, pushed aside the troublesome sovereignty problem (Auburn 1982; Triggs 1986). IGY's cooperative research experience was extended to enable Antarctica's continued treatment as a continent for science and international scientific collaboration. Research data possessing a polar and global utility (e.g. world ozone and pollution levels) prove the region's prime export (Walton and Morris 1990: 265; Walton 1987). Antarctica constitutes a special conservation area subject to strict controls intended to protect its fragile environment against adverse human impacts. In 1991 PREP (Article 2) linked together these aspects to designate Antarctica as 'a natural reserve, devoted to peace and science'.

The shorthand term 'Antarctic Treaty System' (ATS) describes the multilateral arrangements made for the purpose of regulating relations among states involved in Antarctica.[7] PREP defined the ATS as follows:

'Antarctic Treaty System' means the Antarctic Treaty, the measures in effect under that Treaty, its associated separate international instruments in force and the measures in effect under those instruments. (PREP, Article 1f)

In effect, the treaty zone – this covers the area south of 60°S excluding the high seas – was placed under a special intergovernmental regime operating through biennial (annual with effect from 1992) Consultative Meetings (ATCMs) held by rotation in member states and adopting recommendations by consensus. The Antarctic Treaty may last indefinitely; indeed, its preamble mentions the word 'forever' respecting Antarctica's peaceful status. The Treaty may now be reviewed at any time. In fact, the provision for review during or after 1991, that is, once it has been in force for thirty years, led certain commentators to argue mistakenly that the Antarctic Treaty ended in 1991 (Beck 1986: 169–71).

THE FIRST THIRTY YEARS: 1961–91

In 1991 the ATS celebrated its 30th anniversary. During its lifetime the ATS has been characterized by three main trends: the growth in membership, the continuing evolution of the regime's responsibilities and infrastructure, and the wider international community's emerging interest in the region. The 1980s proved a particularly significant period for all three trends, as evidenced by the acceleration in the number of accessions, the conclusion of additional conventions, protocols and recommendations, and the initiation of significant external interest in the region.

First, there has occurred a marked increase in participation. The original 12 signatories, known as the Consultative Parties (ATCPs), have been joined by a further 15 governments adjudged to satisfy the 'substantial research activity' criterion required for this status (Table 7.1). The 26 ATCPs perform the prime decision-making role at ATCMs. A further 14 governments, having acceded to the Treaty and recognized its principles, are merely entitled to observer status at meetings (Table 7.1). In practice, this form of participation is often used as a stepping-stone to ATCP status, given the relatively high cost of establishing Antarctic research and logistical activities. In fact, ATCP criteria have often caused critics to perceive the ATS as a closed, unrepresentative club composed of only a small proportion of the international community, as measured against the 180 plus state membership of the UN. But the regime makes up in quality what it lacks in numbers. Member governments include all the permanent members of the UN Security Council, most European Community states, all the Group of Seven and several major developing states like Argentina, Brazil and India. ATPs account also for over 70 per cent of the world's population.

Second, the regime, though lacking permanent central institutions (e.g. no central secretariat), has evolved in a flexible, pragmatic and cooperative manner designed to accommodate new circumstances and demands as well as to fill perceived gaps through the adoption of ATCM recommendations and the conclusion of additional conventions and protocols on specific issues. Conservation has proved a perennial theme, as evidenced by the 1964 Agreed Measures for the Conservation of Antarctic Fauna and Flora, the 1972 Convention for the Conservation of Antarctic Seals, the numerous recommendations adopted on topics like 'Man's Impact on the Environment', and the adoption of the Protocol on Environmental Protection (October 1991).[8] Resource management has proved another contemporary concern. In 1980 the Convention for the Conservation of Antarctic Marine Living Resources (CCAMLR) was concluded to regulate fishing according to an innovatory ecosystem approach and became effective in 1982.

During the late 1980s it appeared that CCAMLR would soon be paralleled by a minerals regime based upon the Convention on the Regulation of Antarctic Mineral Resource Activities (CRAMRA), as adopted by the ATPs in June 1988 and opened for signature during 1988-9. In the event, strong opposition led by Australia and France resulted in a change of direction. CRAMRA was pushed aside – some believe that it is now effectively dead – by comprehensive environmental protection negotiations, which led to the adoption of PREP and a mining ban (Wolfrum 1991: 1–10). Another interesting institutional trend concerns the extensive linkages (e.g. observer status at ATCMs) formed between the ATS and specialized agencies, like the World Meteorological Organization, the UN Environmental Program (UNEP), and the International Union for the Conservation of Nature and Natural Resources (IUCN).

These inter-connections lead into the third theme concerning Antarctica's emergence as a global question (Beck 1986: 270–9). The ATPs' search for an external accommodation with the wider international community – this is defined to comprise international organizations, non-ATPs, and NGOs – has occurred within the context of changing political, legal, scientific and environmental attitudes. In particular, the Antarctic minerals regime negotiations (1982–8) encouraged many treaty outsiders to seek a share in both the management and benefits of mineral exploitation. The prime forum for debate was the UN, which has conducted annual discussions on Antarctica since 1983 (Beck 1986; Beck 1991c; Beck 1992). The initial consensus approach soon broke down, and since 1985 the international community has been

GLOBAL BOUNDARIES

polarized over the 'Question of Antarctica'. On the one hand, the critics led by Malaysia have pressed for 'collective action under the auspices of the United Nations for an open, accountable and equitable framework for Antarctica for the benefit of mankind as a whole' (Beck 1986: 290–4). Their desire to treat Antarctica as the common heritage of mankind administered by a new UN-based regime influenced by the provisions of UNCLOS has been complemented by a demand for the exclusion of the apartheid regime of South Africa from ATCMs. The UN took up the 'Question of Antarctica' for the ninth successive year during November and December 1991, when the critical lobby pushed through two further resolutions on the topic (UN Press Releases 20 November 1991: 7–10, 6 December 1991: 13–14). One resolution welcomed the recent adoption of PREP, but expressed disappointment that the ATS was developing without the full participation of the international community. Another resolution deplored the continuing participation of South Africa in ATCMs. The 1992 session repeated these themes.

By contrast, the ATPs, albeit prepared to keep non-signatories fully informed of developments, oppose UN interference in the affairs of a region adjudged to be managed by a valid legal regime open to accession by a UN member. Most ATPs, adopting a rather unusual non-participation stance regarding UN discussions and votes on Antarctica, have ignored the resulting resolutions (Beck 1991c: 214–16; Beck 1992: 307). Thus, the two resolutions adopted by the UN General Assembly in December 1991 were passed by large majorities in the face of the non-participation of the ATPs (Table 7.2). One year later, this situation was repeated except that the two themes were merged into one resolution. Critics' attempts to utilize the UN Conference on Environment and Development (UNCED) – the so-called 'Earth Summit' in Brazil during June 1992 – as an alternative avenue for attack have

Table 7.2 UN General Assembly voting on Antarctica resolutions, December 1991

Votes	For	Against	Abstain	Non-participation
Resolution on Protocol on Environmental Protection	65	0	8	43
Resolution on South Africa	73	0	6	38

88

proved abortive. ATPs, though prepared to discuss Antarctica alongside other areas and to provide scientific data, refused to allow the region to be singled out as a special case for treatment (Beck 1991c: 213–14).

UN discussions have been characterized by frequent expressions of support for an agreed approach towards Antarctica, but the international community remains divided concerning the preferred management mechanism. Whether or not agreed answers to the core issues at dispute will emerge in the near future remains questionable, as both sides are seeking a consensus on their own terms. The unbending attitude of the ATPs suggests that the critics will make little progress until they recognize Antarctic and international realities, that is, the need to work for change within, rather than outside, the ATS framework.

ANTARCTIC LAW AND BOUNDARIES

Antarctica has no indigenous population. The presence of only a few thousand transient scientists and support staff is often seen as either obviating any need for the legal apparatus of the modern state or raising only juridical problems of an esoteric character (Fox 1987: 77). However, the legal position of Antarctica remains somewhat tangled, and poses a range of challenges for the lawyer.

A central question concerns the matter of 'Who Owns the Antarctic?' (Luard 1984: 1175). There exists no agreed division of Antarctica, whose 'out of the way' character is qualified by the fact that it amounts to 10 per cent of the world's land surface. Its management raises a range of legal issues – these include interstate conflicts related to jurisdiction, territorial claims, boundaries, resource management and conservation – mirroring international law as a whole. Municipal legal issues are considered in Chapter 9 of this volume.

Antarctica proved the site for the 'last great land rush on earth' (Peterson 1980: 377). Between 1908 and the early 1940s seven governments (Argentina, Australia, Chile, France, New Zealand, Norway, and the United Kingdom) announced claims to most of the continent (Figure 7.2) (Beck 1986: 119–23). Argentinian, British and Chilean claims overlap; indeed, the Antarctica sector between 53°W and 74°W is claimed by all three states. Only one portion of Antarctica (between 90°W and 150°W) remains unclaimed.[9] In fact, this sector must rate as one of the largest, if not the largest, areas on earth left without a claimant. Other ATPs refuse to recognize any claims, while some parties, most notably, the USA and Russia, reserve their rights of sovereignty. The usual focus upon claimants renders it easy to forget that non-claimants make up an equally self-interested grouping anxious to protect their respective rights.

Claims were, and are, justified by a range of factors, including prior discovery, the formal taking of possession, occasional visits by explorers and others, legislative and administrative measures, the sector principle, and permanent 'occupation' through scientific bases. Controversy has centred upon a range of questions. What are the requirements of sovereignty over a polar region? Is a lower standard required than elsewhere? Is it possible for 'effective occupation' ever to be established in the hostile Antarctic environment? How far, if at all, do individual claimants satisfy the requisite criteria? Is Antarctica the common heritage of mankind? Does Antarctica belong to nobody? It is easier to pose questions than to provide agreed answers.

At present, the Antarctic Treaty keeps the lid closed upon a veritable Pandora's box of legal difficulties in which claimants are ranged not only against each other but also against non-claimants and advocates of common heritage. Historically, the sovereignty dispute constituted a serious source of controversy, as outlined earlier, and the resulting search for a conflict avoidance mechanism inspired the negotiations leading to the Antarctic Treaty. Article IV of the latter offered a *modus vivendi* according to which claims were neither renounced, diminished, nor prejudiced, but merely set aside for the duration of the treaty. In effect, the territorial question was evaded through a non-solution freezing individual legal positions. This accommodation has been described as the 'cornerstone' of the Treaty, for Article IV enabled progress on an issue adjudged capable of both causing the failure of the treaty negotiations and thwarting the subsequent development of multilateral cooperation in Antarctica (Auburn 1982: 104; Joyner 1989: 618).

Henceforth, claimants and non-claimants felt able to co-exist within the ATS and develop the regime in a pragmatic, cooperative manner, even if the nature and pace of regime development was largely conditioned by the sovereignty question (Beck 1991d). Resource management, a prime concern for the ATPs since the mid-1970s, assumes the clear identification of ownership for the purposes of regulation (e.g. exploration, exploitation, conservation) and the distribution of benefits. Any Antarctic management scheme encounters two diametrically opposed negotiating positions. Claimants assert that all activity within their respective 'territories' and associated coastal areas, unless specifically exempted (e.g. exchange scientists and observers), is subject to their jurisdiction. Conversely, non-claimants, refusing to recognize the validity of the territorial jurisdiction exercised by claimants claim competence over their own nationals, ships, aircraft, scientific stations and expeditions in Antarctica. If either faction attempted to press its legal stance to a logical conclusion, there could be no progress. In practice, ATPs have been compelled to iron out jurisdictional questions alongside the reconciliation of differing national positions on conservation, exploitation, and other matters. Hitherto sovereignty, though generally depicted as a divisive and destructive factor, has failed to prevent ATPs moving ahead on fundamental topics appertaining to resource exploitation and conservation (Beck 1991d: 255–6).

At present, CRAMRA is generally interpreted as being a dead letter. Perhaps it is, but this much criticized episode is worthy of a more positive interpretation. The ATPs' achievement in negotiating an agreed minerals regime for Antarctica – CRAMRA was adopted by consensus in June 1988 – was qualified initially by the view dismissing the outcome as an abstract solution for a hypothetical problem and then by the manner in which certain ATPs withdrew from the consensus. The resulting controversy distracted attention from the merits of CRAMRA, which was dismissed increasingly – to employ the terminology of Christopher Joyner – as an 'ugly duckling' rather than 'a beautiful swan' (Joyner 1991: 161). In reality, the CRAMRA episode reaffirmed the ATPs' ability to reach an internal accommodation between a range of conflicting interests, even when elaborating a regime regulating mining activities according to strict environmental protection measures and controlling the exercise of sovereignty. CRAMRA, even if it never becomes effective, offers a useful precedent for future legal developments in both Antarctica and the wider world (Wolfrum 1991: 84–94). PREP is normally seen as effectively challenging CRAMRA, most notably, through the imposition of a mining ban, but its text draws heavily on CRAMRA's innovatory provisions regarding say the treatment of adverse human impacts (Wolfrum 1991: 23–31, 84–94).

Existing management regimes, like CCAMLR, provide for a sharing of control over Antarctic space as a whole (Auburn 1984: 273). Whether or not this blurring of sovereignty results in an 'Antarctic territory' – this view has been proposed by Vicuna – is uncertain, especially as sovereignty is deeply rooted (Vicuna 1988: 83). Although territorial sovereignty is being inter-connected in a different and complementary way with the international legal order centred upon the ATS, claims remain a fact of the international politics and law of Antarctica. Nevertheless, Hazel Fox, the Director of the British Institute of International and Comparative Law, has suggested that in future the lawyer might move into a new stage of legal thinking based upon 'an abandonment of old legal methods based narrowly on state sovereignty and territorial jurisdiction in favour of new concepts to accommodate the conflicting demands' (Fox 1987: 78).

The history of the ATS has been marked by a judicious mixture of cooperation and restraint. The regime has managed to move forward because of a willingness by all parties to exercise forbearance, that is, to avoid pushing their respective legal positions to extremes. In the foreseeable future, there seems little prospect of ever securing an agreed answer to the question of 'Who owns Antarctica?'. Hitherto, there has existed merely a consensus to freeze, rather than to resolve, the legal problem. In the meantime, ATPs appreciate that as long as the Antarctic Treaty survives serious legal controversies can be kept just below the surface. Indeed, this feature enhances the prospects for the indefinite duration of the ATS, whose survival and success will prove a function of the parties' support for shared norms concerning the avoidance of points of friction and the promotion of common interests. A freeze on sovereignty, though messy and unsatisfactory on several counts, is perceived as infinitely preferable to any alternative possibility (Peterson 1988: 220–2).

Francis Auburn's accompanying authoritative legal study (Chapter 9) reminds us of the uncertainties surrounding the position of third parties *vis á vis* the ATS. Is the latter an objective legal regime enforceable against other members of the international community? Or does international law establish that a treaty cannot create obligations for a third party without its consent? According to Article X of the Antarctic Treaty, ATPs merely undertake to exert 'appropriate efforts' to prevent activities contrary to its principles. ATPs, opposing external (e.g. UN) interference, allege that Antarctica is managed by a valid, comprehensive and open legal regime embedded in the international political and legal framework and operated for the benefit of the wider international community.

However, most treaty outsiders, influenced by new political and legal attitudes, see things differently. Speaking at the UN First Committee in November 1990, the Zaire delegate epitomized their approach:

> The basic premise that Antarctica is the common heritage of all mankind is enough to explain the major interest of the whole international community in this question ... That continent lies outside the jurisdiction of one country or group of countries and any wish expressed by any country either for its annexation or its exclusive control by a group of countries is nothing but the

expression of an outdated imperialism. (UNGA 1990, A45/CI/ PV42: 18–19, 20 November)

This 'global common' approach, as provided for the deep seabed by the 1982 UN Convention on the Law of the Sea (UNCLOS), confronts head-on both the ATS' claim to 'govern' Antarctica and the principle of territorial sovereignty advanced by the seven claimants. The debate continues. Related areas of uncertainty include maritime boundaries in the Southern Ocean, the relationship between the ATS and UNCLOS, and the legal status of ice (Joyner 1989: 613).[10] Another contemporary problem concerns the intentions of one prominent UN critic of the ATS, Pakistan, which became active in Antarctica during 1990–1 without acceding to the Antarctic Treaty (*The Muslim* 1991).

CONCLUSION

Antarctica's contemporary visibility is only relative, and should not blind us to the region's *perceived* fundamental insignificance. But we should not overlook the continuing capacity of allegedly insignificant and ignored parts of the globe to move to centre stage with surprising consequences. It is unwise to dismiss Antarctica as either a pole apart, a wasteland or a mere symbol. Human activities conducted in Antarctica are capable of exerting significant impacts throughout the globe in terms of climate, ocean currents and world sea levels. Any substantial melting of the Antarctic ice sheet – this covers most of the continent often to a depth of 4 miles – would exert severe impacts. Few countries would remain unaffected, while the resulting rise in world sea levels would pose the ultimate threat to the survival of low-lying states, like the Maldives and Bangladesh.

Antarctica has become part of the currency of environmentalism. Although politicians appear to be talking the same language of conservation, as demonstrated by the emphasis upon the 'ozone hole', 'CFCs', global warming and so on, the subject of Antarctica at the UN debates, as in the CRAMRA and PREP episodes, has highlighted already the potential for disagreement, even conflict, at both the national and international levels about appropriate strategies to secure agreed objectives. Environmental issues, though often regarded as transcending politics, raise fundamental political principles (Newby 1989). In turn, legal concepts, like sovereignty and common heritage, spill over into politics. The ATS' future course will prove ultimately a matter of political will, that is, a function of the individual and collective interests of ATPs as well as of their capacity to uphold, and to be seen upholding, the interests of the wider international community, especially on environmental matters.

Thirty years on the ATS is interpreted still by ATPs as the most appropriate framework for the pursuit of their respective Antarctic objectives, and particularly for avoiding the unmanageable escalation of national policy interests at variance with each other (Brook 1984: 256). However, non-ATPs advance a less flattering view of the ATS, thereby explaining the divide currently existing in the international community regarding – to quote the title of the UN agenda topic – the 'Question of Antarctica'. In addition, the ATPs, though endeavouring to promote multilateral cooperation between an ever-increasing number of states on a widening range of responsibilities, face management problems of growing complexity. Past achievements do not guarantee a trouble-free future. The most pressing topic is the implementation of PREP, which provides a framework for safeguarding the Antarctic environment from adverse impacts consequent upon human activities. Tourism, albeit not completely unregulated, is an urgent case for action, perhaps through either an additional annex to PREP or recommendation (Beck 1990a: 354; Final Report of XVIth Antarctic Treaty Consultative Meeting, 1991: 25).

The global environment has moved towards the top of the political agenda. During May 1990 the Bergen Conference took up the con-servationist and sustainable development themes outlined by the 1987 Brundtland Report, which included an examination of Antarctica's future in the light of a growing awareness of both the 'unprecedented' pressures on the global environment and the value of international cooperation to pursue common goals (Brundtland Report 1987: 12–13).

> New questions about equitable management are presenting challenges that may re-shape the political context of the next decade. During the forthcoming period of change, the challenge is to ensure that Antarctica is managed in the interests of all human-kind, in a manner that conserves its unique environment, preserves its value for scientific research, and retains its character as a demilitarised, non-nuclear zone of peace. (275)

In the near future, it seems difficult to anticipate the achievement of a consensus on the part of the whole international community regarding the 'equitable management' of the 'last great wilderness on earth'. At present, ATPs and non-ATPs provide conflicting answers to these 'new questions'.

NOTES

1 Declaration by ATPs on the 30th anniversary of entry into force of the Antarctic Treaty, in the *Final Report of XVIth Antarctic Treaty Consultative Meeting*, Bonn, 18 October 1991.
2 UNCLOS, though having 158 signatories, has yet to secure the requisite number (60) of ratifications to become effective. By October 1991, UNCLOS had been ratified by 50 states: *UN Press Release*, 13 November 1991.
3 See also interview with the then Prime Minister of the UK, Margaret Thatcher, BBC Television, *Nature Programme*, 2 March 1989.
4 There is as yet no generally accepted acronym for this protocol.
5 Whaling is covered by the 1946 International Convention for the Regulation of Whaling.
6 Note the work of Professor Bernard Herber of the University of Arizona at Tucson, Professor Kenneth White of the University of British Columbia, Vancouver, is another economist working on the subject.
7 The creation of a secretariat has been under discussion for several years. The November 1992 Antarctic Treaty Consultative Meeting held at Venice showed signs of progress on the topic.
8 PREP, though adopted in October 1991, has yet to come into effect. However, the 1991 ATCM agreed that, pending PREP's ratification, its provisions should be applied by parties.
9 This unclaimed sector is relatively inaccessible (e.g. because of ice) and economically unattractive, although at one stage the USA showed signs of interest.
10 Also refer to Joyner 1988, 1990; Safronchuk 1991: 328–33; and Mangone 1988).

REFERENCES

Auburn, F.M. (1982) *Antarctic Law and Politics*, London: Hurst.
—— (1984) 'The Antarctic minerals regime: sovereignty, exploration, institutions, and environment', in S. Harris (ed.) *Australia's Antarctic Policy Options*, Canberra: CRES.
Australian House of Representatives Standing Committee on Environment, Recreation and the Arts (1990) *Tourism in Antarctica*, Canberra: AGPS.
Beck, P. J. (1986) *Antarctica in International Politics*, London: Croom Helm.
—— (1987) 'A cold war: Britain, Argentina and Antarctica', *History Today* 37, 6.
—— (1988a) *The Falkland Islands as an International Problem*, London: Routledge.
—— (1988b) 'A continent surrounded by advice: recent reports on Antarctica', *Polar Record* 24, 151: 285–91.
—— (1989) 'British relations with Latin America: the Antarctic dimension', in V. Bulmer-Thomas (ed.) *Britain and Latin America: A Changing Relationship*, Cambridge: Cambridge University Press.
—— (1990a) 'Regulating one of the last tourism frontiers: Antarctica', *Applied Geography* 10, 4: 343–5.

—— (1990b) 'Antarctica as a zone of peace: a strategic irrelevance? A historical and contemporary survey', in R.A. Herr, H.R. Hall and M.G. Haward (eds) *Antarctica's Future: Continuity or Change?*, Hobart: AIIA/Tasmanian Government Printing Office: 193–220.

—— (1991a) 'The conflict potential of the "dots on the map"', *The International History Review* XIII, 1: 124–33.

—— (1991b) *Why Study Antarctica?*, Kingston-upon-Thames: Apex Centre, Kingston Polytechnic (now Kingston University).

—— (1991c) 'Antarctica, Viña del Mar and the 1990 UN debate', *Polar Record* 27, 162: 211–16.

—— (1991d) 'The Antarctic resource conventions implemented: consequences for the sovereignty issue' in A. Jorgensen-Dahl and W. Ostreng (eds) *The Antarctic Treaty System in World Politics*, London: Macmillan: 240–51.

—— (1992) 'The 1991 UN session: the environmental protocol fails to satisfy the Antarctic Treaty System's critics', *Polar Record* 28, 167: 307–14.

Beeby, C. (1991) 'The Antarctic Treaty System: goals, performance and impact' in A. Jorgensen-Dahl and W. Ostreng (eds) *The Antarctic Treaty System in World Politics*, London: Macmillan: 193–220.

Brook, J. (1984) 'Australia's policies towards Antarctica', in S. Harris (ed.) *Australia's Antarctic Policy Options*, Canberra: CRES.

Bruntland Report (1987) Newark: United Nations.

Cousteau, J. (1990) *Hearing of the Sub-Committee on Oceanography and Great Lakes, Committee of Merchant Marine and Fisheries, House of Representatives*, USA, 2 May: 4.

Department of the Arts, Sport, the Environment, Tourism and the Territories (DASETT) (1988) *Memorandum on Tourism in the Antarctic*, Australia, 6 October.

Eco (1990) LXXVII, 1, 19 November: 1.

Enzenbacher, D. J. (1992) 'Tourism in Antarctica: numbers and trends', *Polar Record* 28, 164: 17–20.

Fox, H. (1987) 'The relevance of Antarctica to the lawyer', in G.D. Triggs (ed.) *The Antarctic Treaty Regime: Law, Environment and Resources*, Cambridge: Cambridge University Press.

Harris, S. (ed.) (1984) *Australia's Antarctic Policy Options*, Canberra: CRES.

Herr, R.A., Hall, H.R. and Haward, M.G. (eds) (1990) *Antarctica's Future: Continuity or Change?*, Hobart: AIIA/Tasmanian Government Printing Office.

James, A. (1984) 'Sovereignty: ground rule or gibberish?', *Review of International Studies* 10, 1: 16–17.

Jorgensen-Dahl, A. and Ostreng, W. (eds) (1991) *The Antarctic Treaty System in World Politics*, London: Macmillan.

Joyner, C. C. (1988) 'The Antarctic legal regime and the law of the sea', *Oceanus* 31, 2: 22–7.

—— (1989) 'The evolving Antarctic legal regime: review article', *American Journal of International Law* 83, 3: 618.

—— (1990) 'Maritime zones in the southern ocean: problems concerning the correspondence of natural and legal maritime zones', *Applied Geography* 10, 4: 307–25.

—— (1991) 'CRAMRA: the ugly duckling of the Antarctic Treaty System?'

in A. Jorgensen-Dahl and W. Ostreng (eds) *The Antarctic Treaty System in World Politics*, London: Macmillan: 161.

Larminie, F.G. (1987) 'Mineral resources: commercial prospects for Antarctic minerals' in G.D. Triggs (ed.) *The Antarctic Treaty Regime: Law, Environment and Resources*, Cambridge: Cambridge University Press: 176–81.

Luard, E. (1984) 'Who owns the Antarctic?', *Foreign Affairs* 62, 5: 1175.

Mangone, G. (1988) 'The legal status of ice in international law', in R. Wolfrum (ed.) *Antarctic Challenge III*, Berlin: Duncker and Humblot: 371–88.

The Muslim (1991) 'Hoisting the Pakistani Flag in Antarctica', 11 January, 22 February.

Newby, H. (1989) *Ecology, Amenity and Society: Social Science and Environmental Change*, Patrick Abercrombie Memorial Lecture, University of Liverpool, 31 October.

Parsons, A. (1987) *Antarctica: the Next Decade*, Cambridge: Cambridge University Press.

Peterson, M.J. (1980) 'Antarctica: the last great land rush on earth', *International Organization* 34, 3: 377.

—— (1988) *Managing the Frozen South: the Creation and Evolution of the Antarctic System*, Berkeley: University of California.

Quigg, P. (1983) *A Pole Apart*, New York: McGraw Hill.

Safronchuk, V. (1991) 'The relationship between the ATS and the Law of the Sea Convention of 1982' in A. Jorgensen-Dahl and W. Ostreng (eds) *The Antarctic Treaty System in World Politics*, London: Macmillan.

Shackleton, L. (1982) *Falkland Islands Economic Study*, London: HMSO.

Sugden, D. (1982) *Arctic and Antarctic: A Modern Geographical Synthesis*, Oxford: Blackwell.

Triggs, G.D. (1986) *International Law and Australian Sovereignty in Antarctica*, Sydney: Legal Books Pty.

—— (ed.) (1987) *The Antarctic Treaty Regime: Law, Environment and Resources*, Cambridge: Cambridge University Press.

Vicuna, F.O. (1988) *Antarctic Mineral Exploitation: The Emerging Legal Framework*, Cambridge: Cambridge University Press.

Walton, D. (ed.) (1987) *Antarctic Science*, Cambridge: Cambridge University Press.

Walton, D. and Morris, E. (1990) 'Science, environment and resources', *Applied Geography* 10, 4: 265–86.

Willan, R., MacDonald, D. and Drewry, D. (1990) 'The mineral resource potential of Antarctica: geological realities', in G. Cook (ed.) *The Future of Antarctica: Exploitation Versus Preservation*, Manchester: Manchester University Press: 25–52.

Wit, Maarten de (1985) *Minerals and Mining in Antarctica: Science and Technology, Economic and Politics*, Oxford: Oxford University Press.

Wright, N.A. and Williams, P.L. (eds) (1974) *Mineral Resources of Antarctica*, Washington DC: US Geological Survey.

Wolfrum, R. (ed.) (1988) *Antarctic Challenge III*, Berlin: Duncker and Humblot.

—— (1991) *The Convention on the Regulation of Antarctic Mineral Resource Activities*, Berlin: Springer Verlag.

8

THE FUTURE OF THE ANTARCTIC TREATY – THIRTY YEARS ON

Hernan Santis Arenas

INTRODUCTION

After thirty years of the operation of the Antarctic Treaty (1959), it is apposite to explore the horizons in relation to the juridico-political development of the 'Antarctic System'. It is also useful to identify some new themes that could agitate international relations in connection with Antarctica.

By the early 1950s, the Antarctic Continent was converted into a potential area of conflict for some of the states interested in the austral polar lands. For example, from 1940, the moment Chile proceeded to delimitate the Chilean Antarctic Territory, its relations with the United Kingdom and the Argentinian Republic faded. The Antarctic claim of Argentina in 1943 also darkened the relations between these three actors. From the end of the Second World War, relations between these three states, because of their Antarctic aspirations, passed rapidly towards potential juridico-politico-territorial conflict. Argentina and Chile, in 1947 and 1948, recognized reciprocally their historical rights to the territories in Antarctica. The United Kingdom hardened its political position. On many occasions tensions threatened to spill over into acts of aggression.

The Great Powers, the USA and the USSR, in accordance with their objectives of world hegemony, evaluated diplomatic difficulties in relation to Antarctica in many different ways. The USA and the USSR did not have expressly stated territorial aspirations. They were interested only in the scientific investigation in relation to the territory and in the experimentation of many technologies and techniques in the polar environment. In this context, the USA tried to promote and encourage better diplomatic understanding so as to avoid a potentially serious conflict involving some of its European and Latin American allies. International Geophysical Year (1956–7) facilitated the cross-exchange of views of hundreds of scientists engaged in different research programmes. It also encouraged the view that we need sufficient time to augment our collective knowledge of the 'white continent'.

As a researcher of political and spatial processes my aim in this chapter is to focus on specific issues in the development of the Antarctic Treaty that are of politico-geographical relevance. The chapter examines the thirty years from 1961 and 1991 in order to highlight certain aspects of state-thinking that have not been given so much attention by other scholars. Finally, the chapter will examine some possible future sources of tension.

DEVELOPMENTS IN THE ANTARCTIC TREATY

It is not necessary here to delve too much into the historical developments of the Antarctic Treaty, because they are sufficiently known (see for example, various works by Peter Beck, including Chapter 7 in this volume) and there are numerous scientific publications that have taught us how much is still unknown about the white continent. Nevertheless, new technologies such as satellite images have amplified the geological, mineralogical, glaciological, oceanographic and meteorological knowledge of the structure and natural processes in the Antarctic polar region.

From another perspective, scientists in fields such as biology, zoology, botany and chemistry have revealed aspects of the macro-organic and micro-organic world. Geophysicists, climatologists and meteorologists are now carefully analysing the depletions in the ozone layer.

Three decades of scientific cooperation

No researcher whether from the sciences or social sciences, can ignore the fact that international scientific cooperation helps to explain in a great measure the fact that twelve governments (Argentina, Australia, Belgium, Chile, France, Japan, New Zealand, Norway, Union of South Africa, the then USSR, UK and the USA) have shelved (but not removed) their territorial, juridical and political disputes in relation to Antarctica or sections of it.

Governments expressly agree to use Antarctica only for pacific objectives, forbidding all measures of a military character, which has helped establish a conducive research environment. The activities of the

Scientific Committee on Antarctic Research (SCAR) and the contributions of national and governmental scientific groups from the original signatory countries and from countries who later became signatories show that international scientific cooperation has effectively materialized.

This is evident in the laborious efforts to postpone for fifty more years (until 2041) discussion about the exploitation of mineral resources. Meetings at Viña del Mar (Chile) and Madrid (Spain) show that the states with Antarctic interests can be classified into three groups. Some prefer that the continent and the adjacent seas, down to the parallel 60°S should be declared definitely as a reserve and natural laboratory; others prefer to accept selective, technically-controlled exploitation of certain live and mineral resources; others support the opening up of the continent to economic exploitation on a great scale.

From an individual and scientific perspective, not forgetting that I am also a taxpayer of one of the states with Antarctic interests, it must be recognized that we cannot take definitive decisions about Antarctica without first evaluating the potential environmental, social and political impacts of economic exploitation. It is in the interests of all to preserve the fragile Antarctic ecosystem whilst learning more about it. International scientific cooperation does allow for theoretical and utilization research to be carried out, and indeed, this is one of the merits of the Antarctic Treaty, but it is insufficient on its own if the scientists themselves do not have the capability to influence their respective governments in the direction of political and economic cooperation to safeguard the future of Antarctica. This is an aspect of the future potential 'political tensions' which will be discussed later.

Building a new economic approach: Resource use

Some scientists have been worried about the economic objective in Antarctic research. There is tension between the eagerness for a rational knowledge of Antarctica and the economic and material purposes of the national societies to which we belong. Unfortunately, nationalist exaltations and the imperial and strategic concerns of nation-states have introduced the idea that Antarctica is a new kind of 'El Dorado', with huge resource potential. The objectives of owning lands there and of exploiting its resources have motivated many members of the political and military élites of states.

A mixture of reasons helped to prevent the various nation-states with Antarctic interests from advancing projects to exploit the marine riches and Antarctic terrestrial space. These included lack of adequate technology and financial resources to support such projects. The authors and editors of the Antarctic Treaty were careful to establish a true security mechanism to avoid misuse and damage to the Antarctic environment. This helped establish pacific use of the territory, including the notion of protecting and conserving its live resources, which, in the consulting sessions, gave birth to the protection and conservation of the whole environment. The signatory countries of the Antarctic Treaty, at least till now, have never been opposed to resource exploitation, but they prefer adequate measures for the conservation of the physical environment. They have resolved in the majority of cases to postpone the exploitation of the natural resources until later.

The signatory states have also examined how future exploitation of resources should be organized. Some think a multinational entity should be created, whose members – the same signatory states – should manage economic exploitation. Others are in favour of facilitating concessions to national, multinational and transnational private business. So the Treaty did not rule out future exploitation. Rather, it allows for a new economic approach, which can only become operable with better technology and effective procedures to guarantee the protection and conservation of the environment.

Membership increase

The original twelve states with Antarctic interests have now risen to forty (refer to Table 7.1, p. 77). The present signatory states can be divided between those with territorial interests and those without. Each state exhibits its own motives for its interest in Antarctica, including historical, juridical, scientific, political, and economic reasons. Some of these states may be interested in 'being there' at some future time to participate in the division of the continent. It is useful to consider the territorial and other ambitions of the states in order to appreciate possible future tensions concerning Antarctica.

POLITICO-GEOGRAPHICAL EVOLUTION: FROM TERRITORIAL CLAIMS TO INTERNATIONAL PROPERTY

It is worth examining the politico-geographical evolution of the interests in Antarctica and the Antarctic Treaty (Figure 8.1).

During 1908 the British government decided to extend its territorial

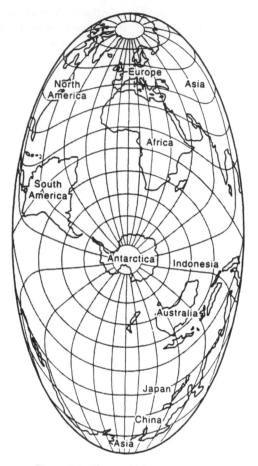

Figure 8.1 The world from Antarctica

sovereignty from the Falkland Islands (Malvinas Islands to Argentina) to Antarctica. The meridians 20° and 80° and the parallel 50°S gave the bounds to the lands and seas of the new Antarctic Dependency. It is certain too that in those days, there were few Chilean and Argentinian citizens in that delimitation. Over the next decades, with different arguments attached to the British claim, other governments communicated their own territorial claims (New Zealand 1923; France, 1924; Australia, 1933; Norway, 1939; Chile, 1940; Argentina, 1943) (see Figure 7.2 p. 84).

As Kidron and Segal (1982: 4) point out the nation-state began its invasion of Antarctica in the early 1900s and it eventually culminated in the signature and ratification of the Washington Treaty or Antarctic Treaty of 1959. Gradually, the continent and the seas covered by the parallel 60°S came to be regarded as the geographic area under the co-administration of the signatory states of the Treaty. The signatory states began to regulate many different aspects of their territorial relations in Antarctica. Between 1961 and 1991 states interested in scientific research of Antarctica's seas and land, if they could meet the norms and regulations agreed to by the signatory states, could also become signatories. The Antarctic Treaty helped to avoid any controversies or conflicts over territory by introducing the *status quo* in territorial matters.

After thirty years of the Treaty, it is possible to detect several different attitudes towards Antarctic territory and resources. Some governments became supporters of the indefinite maintenance of the *status quo* in territorial matters and continuing to allow the entrance of new states under the conditions that prevailed. Another group argues that Antarctica must be indefinitely conserved as a natural laboratory. The rest argue for joint ventures between the signatory states to exploit resources under certain conservation and protection regulations and guidelines.

If an end to the *status quo* was sought, it is clear that to avoid discrepancies, controversies and territorial conflicts, it would be necessary to agree on mechanisms to forestall the ambitions of the states with territorial claims (Argentina, Australia, Chile, France, Norway, New Zealand, the UK) and facilitate the access of the rest.

The group of states seeking ways to conserve the white continent for all humankind and for future generations can be further subdivided into those states favouring some kind of natural reserve for coming generations, and those perferring the internationalization of the continent under the aegis of a world authority, such as the United Nations, to keep vigil over the exploitation of organic and inorganic resources and distribution of those resources.

For a variety of reasons associated with the development of the Antarctic Treaty there has been an evolution away from competing territorial claims towards concepts of co-administration and a multi-state property or international property.

NEW DEBATES

There are four key areas of debate. These concern mining development, the issue of keeping Antarctica as pollution-free as possible, issues of what kind of nation-state (or group of nation-states) property or world property it should be, and the revision of the Antarctic Treaty.

Minerals development

The consulting meeting in Madrid (1991) agreed to postpone for fifty years the exploitation of Antarctic mineral resources, so that future generations in 2041 will be those making the decisions. It seems appropriate to explore the significance of this development further.

Interpretations of the results of geological and economico-geological research, often under the influence of nationalist or internationalist political ideas, have resulted in hasty conclusions that the area's geological formations contain 'immensurable mineral resource'. The very fact that inventories of mineral resources located in different Antarctic areas, such as chrome, nickel, cobalt, copper, gold, silver, manganese, molibdene, iron, titanium, platinum, lead, zinc, tin and uranium, plus coal, oil and natural gas deposits, have been drawn up is bound to awaken the economic interests of many companies and states. But the fact is that these minerals are not economically exploitable with current financial and technological resources. What is clear is that the perceptions of different states towards Antarctic resources may change fundamentally over the coming decades, particularly in advanced industrial economies that are unable to conserve their own energy and mineral resources or to curb the demands of domestic consumers.

In 1992 there was the celebration of 500 years of the so-called 'discovery' of the New World. The first conquerors of the new lands opened the doors for the migration of millions of Mediterranean and Atlantic Europeans. The expansion of Europe and of the European concepts of the nation-state and territorial sovereignty have transformed the world political map into one made up of sovereign state territories divided by precise international boundaries. With the end of the Cold War we live in a multi-polar world, but one which is characterized by much uncertainty, particularly with regard to international order and security. We also live in a world in which geopolitical and geoeconomic concerns are increasingly intertwined. Over the thirty years, 1961–91, the number of Antarctic Treaty adherents has grown rapidly, and it will be surprising if in the next three decades more states do not become signatories to the Treaty. Under circumstances of political and economic insecurity in the world interstate system it is difficult to be entirely optimistic regarding the attitudes of some of these states and the older signatories towards the resources of Antarctica.

An area free of pollution?

The deepening understanding of science concerning the structures and processes of the environment have taught us that natural inter-connections on a planetary scale exist and function. Many scientists, intellectuals, politicians and ordinary inhabitants of the planet consider it a duty to press for protection and conservation programmes in order to maintain these earth systems for future humankind. Perhaps this explains in great measure the implicit conservationist and protectionist attitudes in the Antarctic Treaty.

From the consultative meetings of Viña del Mar and Madrid we can see that some states wish to keep the ecological systems of Antarctica free from economic exploitation of both the marine and land environments. Other states argue for resource exploitation subject to stringent rules and guidelines. Recent scientific research has provided much data to support the view that highly localized uses of resources can have serious ramifications in terms of soil, air and water pollution well beyond the areal extent of the resource exploitation.

Indeed, even without direct exploitation within the white continent there are serious concerns about such matters as ozone depletion and damage to both Antarctica and large parts of the planetary systems.

There are many pollution issues to consider if any mining activity is ever to be permitted in Antarctica. For example, there are no definite scientific conclusions concerning the diffusion of arsenic in the refining processes that permit the production of millions of tons of copper each year. It is unlikely that the producer countries and exporters of this metal would finance investigations in Antarctica, and yet they include signatory states of the Antarctic Treaty.

Nation-state property or the property of humankind?

There are some fascinating international political relations issues relating to the viewpoints of the South American states with Antarctic interests, and perhaps of other Antarctic Treaty signatory states. Argentina and Chile are both signatory states with territorial claims. Their respective governments often allude to their 'Antarctic rights', which they back up with various juridical, historical, geographical and political arguments. There is little evidence in documents to suggest that these states have changed their positions, although they maintain the status quo in order to facilitate bilateral relations. In 1985/6, following the Treaty of Peace and Friendship between these states to resolve a maritime delimitation dispute, both sides excluded 'the Antarctic matter'.

The situation has become more complex during the last thirty years with states like Brazil (1975), Uruguay (1980), Peru (1981), and Ecuador (1987) with different juridical and geographical perspectives, becoming incorporated into the group of states with Antarctic interests. Brazil and Peru have developed their territorial interests and claimed their so-called 'Antarctic rights'. As such, it is possible to envisage new tensions in future relations between the South American states of Brazil, Argentina, Peru and Chile over their respective 'interests' and actual claims and over their broader geopolitical aspirations which incorporate Antarctica. In fact the territorial interests of Brazil and Peru overlap those of Argentina and Chile. The modern history of humankind and past territorial disputes between South American neighbours makes it important not to underestimate the territorial aspirations of states, whatever their validity under international law.

Re-examination of the Antarctic Treaty

Several scholars have examined the Antarctic Treaty (1961) and suggested modifications. My focus is on the clauses of the Treaty which support the territorial status quo. There seem to be the following options. First, maintenance of the Antarctic System in which states are involved in the territorial co-administration of the continent according to basic principles and objectives laid down by the Treaty. Second, modifications to the juridical system in which Antarctica becomes the common property of humankind. This would, however, need binding international agreements, and it would probably involve an indefinite prohibition of mineral resource exploitation in Antarctica.

The Antarctic Treaty has worked well. It is clear that the emphasis on the *status quo* has not eradicated territorial claims, but it has kept them in check and avoided conflict. The Treaty has impeded the irrational exploitation of Antarctic resources, and it has allowed a system of international cooperation to function, no matter how imperfect that may be.

TENSIONS ON THE HORIZON

It is worthwhile to identify some possible future tensions concerning Antarctica. The following points are derived largely from a South American perspective.

Territorial claims

In 1958, the Brazilian government formally protested to the USA because it had not been invited to participate in the Antarctic conference. In the protest it was stated that Brazil was obliged to protect its national security and keep the right of access to Antarctica. Some years later, Brazil fulfilled the formalities of becoming an adherent to the Antarctic Treaty on 16 March 1975.

It is interesting here to mention the influential geopolitical thesis dveloped by Couto e Silva (1967) and T. de Castro (1976). The former appraises Antarctica in terms of Brazilian and South American security interests, and the latter developed the theory of '*defrontacáo*' (the fronting of coasts) through the extreme meridians of the coasts of Brazil, Ecuador, Peru, Chile, Argentina and Uruguay (Figure 8.2). The Argentinian geopolitical literature has taken up these themes and left the door open to possible future controversy and disputes between Argentina and Brazil, and Chile and Brazil.

In 1979, a new Peruvian political constitution was elaborated, and it should be noted that it included mention of Antarctic rights based on geographic projections from her coastline, historical factors and the state's right to participate in the discussions of Antarctica's future. In 1981, Peru applied to become an adherent to the Antarctic Treaty and in 1983 it created the Antarctic National Commission. An academic institution, the Peruvian Institute of Geopolitical and Strategic Studies was responsible for the circulation of a collection of conference papers under the heading 'Peru and the Antarctic' in November 1984. In one of those papers it was suggested that 'fronting of coasts' would form the basis of future Peruvian territorial claims and arguments relating to Antarctica. Lopetegui (1986) argues that these new Peruvian aspirations imply claims that overlap the territory included in the 1940 delimitation of Chile.

Of course, the very fact that Brazil, and Peru, together with Uruguay and Ecuador have adhered to the Antarctic Treaty, suggests that the above discussion concerning possible discrepancies and controversies between the South American states is exaggerated. But it should not be overlooked that in South American societies, particularly in political (military) circles, the notions of territoriality, the territorial state, and territorial sovereignty are strong.These notions are deeply rooted in the intelligentsia and political élites. While the territorial status quo remains in force under the Antarctic Treaty with respect to various claims of sectors of the continent, there is unlikely to be any problem or conflicts,

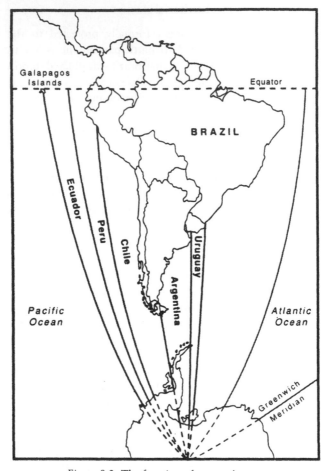

Figure 8.2 The fronting of coasts theory

but once the juridical conditions change, there could be renewed tension over territorial matters.

Historical claims and neighbouring states

Before the Antarctic Treaty, particularly in the 1940s and 1950s, Chile and Argentina had recognized their mutual 'Antarctic territorial rights' (1947 and 1948) and were locked in confrontation with Britain. Indeed, this potentially serious confrontation seems to have been an important factor in leading the USA to seek an agreement.

THE FUTURE OF THE ANTARCTIC TREATY

Whilst the Antarctic Treaty has maintained a *status quo* it has not removed the territorial question altogether. In the Peace and Friendship Treaty between Argentina and Chile concerning maritime delimitation in the austral seas (Santis Arenas 1990), both states explicitly affirmed that the agreement between them could not be invoked in relation to their respective Antarctic claims which of course, overlap each other.

It is important to consider the broader context encompassing the relations these two states have with each other and with other neighbours. Argentina and Brazil dispute their respective economic influence over the Rio de la Plata basin. They have also been wary of each other's influence and presence in the American Antarctic sector. Chilean and Peruvian boundary and territorial problems are related mostly to the efforts of some intellectuals and politicians to claim areas of territory that were the object of treaties of peace and friendship and of territorial limits in 1883 and 1929 respectively. Both states have ambitions to increase their respective influence and domains in the south-eastern Pacific Ocean. Peruvian Antarctic territorial pretensions are oriented directly over the sectoral Chilean delimitations of 1940.

In recent times many Chilean and Argentinian authors, including some members of the political élites of these two countries, have expressed the desire that their respective governments should work closely together over Antarctic affairs. It is argued that it is in their mutual interest to have better intergovernmental cooperation to improve their position within the Antarctic juridical system.

Whilst the situation in the early 1990s has been one of gentle diplomacy and a more cooperative spirit has been fostered by the Antarctic Treaty, everything could change in the not too distant future. Modifications to the treaty arrangements that do nothing to satisfy the territorial aspirations of these South American countries may open the way for renewed dispute, even conflict. The unfortunate fact is that the territorial factor cannot be so easily discounted.

Geostrategic objectives

During the Second World War, some of the maritime powers involved in the conflict were very interested in occupying areas opposite Cape Horn. The motives for this were to protect their own vessels and to control the Atlantic–Pacific interoceanic maritime traffic.

The maritime traffic through the Strait of Magellan has tonnage size and type of load restrictions on it which do not apply to the Drake Pass. Even so, the protection and control of these maritime routes which separate South America from Antarctica are a geostrategic concern. These maritime concerns may become even more important in the twenty-first century, particularly in the event of a generalized conflict affecting that part of the world.

Differences between rich and poor countries

Cerda (1988) has analysed relations between rich and poor countries and the influence these relations have on Antarctic interest. Some of the most critical states regarding the Antarctic regime are poorer developing countries, and some of these states are also signatories to the UN Convention on the Law of the Sea (UNCLOS III 1982). For these countries, the possibility of resource wealth in Antarctica, plus their economic difficulties at home, may lead to intensifying pressure in future to seek ways to gain access to some of Antarctica's resources.

CONCLUSION: CHANGING POLITICAL ORDER

There have been sudden and sweeping changes to the international political order since 1988. The end of the Cold War has led to a new and more complex picture. At its simplest, the world appears to be increasingly 'tripolar' in terms of political and economic power between North America, Japan and Europe (the European Community). But this does not necessarily mean we will be living in a more stable world system. Furthermore, there are many questions relating to how the changing sets of political, military and economic power relationships will affect the Antarctic interests of various states. From the perspective of South America it is clear that we should not dismiss the possibility of future territorial disputes and resource disputes over Antarctica.

REFERENCES

Berguño, J. (1987) 'Realidad y régimen jurídico y político de la Antárctica', *Revista Chilena de Geopolítica*, 4, 1: 49–58.
Cerda, J. (1988) 'Análisis Geopolítico de la cuestión Antártica, La Provincia como vértice del espacio poliítico nacional', Santiago: Uinversidad de Chile, Instituto de Ciencia Política.
De Castro, T. (1976) *Rumo á Antártica*, Rió de Janeiro: Livraria Freitas Bastos.
Kidron, M. and Segal, R. (1982) *The State of the World Atlas*, London: Pluto Press.
Lopetegui, T. (1986) *Antártica un desafío perentorio*, Santiago: Instituto Geopolítico de Chile.

Mercado, J.E. *et al.* (1984) *El Perú y la Antártica*, Lima: Instituto Peruano de Estudios Geopolíticos y Estratégicos.
Neal, V.T. (1987) 'Cooperación científica internacional en la Antártica', *Revista Chilena de Geopolítica*, 4, 1: 153–6.
Palma, G.S. and Mujica, R. (1987) 'Recursos antárticos para el desarrollo',*Revista Chilena de Geopolítica*, 4, 1: 129–51.
Santis Arenas, H. (1987) 'Visión geopolítica del Cono Sur de América', *Revista Chilena de Geopolítica* 4, 1: 5–19.
—— (1987) 'Importancia geopolítica de la Antártica', *Revista Chilena de Geopolítica* 4, 1: 101–14.
—— (1990) 'The nature of maritime boundary conflict resolution between Chile and Argentina, 1984', in C. Grundy-Warr (ed.) *International Boundaries and Boundary Conflict Resolution*, Durham: Boundaries Research Press.
Santis Arenas, H. and Riesco, R. (1987) *Las fronteras Antártica de Chile*, Santiago: Uinversidad de Chile, Instituto de Ciencia Política.

9

THE ANTARCTIC LEGAL SYSTEM

Francis Auburn

THE ANTARCTIC SYSTEM

This chapter will examine some of the legal issues arising under the Antarctic System. The Treaty permits a Review Conference after the expiration of thirty years; this took place in 1991 (Article XII (2)).[1] This formal procedure has been seen as dangerous because it provides that the amendment may be adopted by a majority of the Contracting Parties, which must also include a majority of the Consultative Parties. If this has not been done within two years, any Contracting Party may give two year's notice of withdrawal.

In practice, amendments of the Treaty have usually been accomplished by indirect means. One means is re-interpretation of the accepted meaning of the Treaty, for example, extension of the Treaty to cover the high seas (Auburn 1982: 130). Another means is by separate agreements which effectively alter the Treaty. So the Convention on the Regulation of Antarctic Mineral Resource Activities (CRAMRA) (Auburn 1990: 259) would have effectively amended the Treaty.[2]

The Protocol on Environmental Protection states that it shall 'supplement the Antarctic Treaty and shall neither modify nor amend that Treaty' (Article 3(1) Protocol to the Antarctic Treaty on Environmental Protection (29 April 1991) XI ATSCM/2/30), but the Protocol would effectively alter the Treaty. For instance, it is provided that inspections shall be carried out in accordance with Article VII of the Treaty to promote the protection of the Antarctic environment and dependent and associated ecosystems and to ensure compliance with the Protocol (*Ibid.* Article 13(1)). But inspection under Article VII of the Treaty is to be carried out 'in order to promote the objectives and ensure the observance of the present Treaty' (Article VII(1) Antarctic Treaty). The only hint of environmental protection in the Treaty is the reference to

112

'preservation and conservation of living resources in Antarctica' as a subject for measures in furtherance of the Treaty (*Ibid.* Article IX (1) (f)). Inspection under the Protocol can be seen only as a desirable aim and it is not the present purpose to criticize the substance of the Protocol on this point, but it is intended to point to the rigidity of the Treaty and the legal difficulties involved in creating regimes which purport to preserve the Treaty unaltered, whilst in practice amending it. Legal fictions often have a creative role to play, but in this instance the fundamental problems of the legal regime created by the Treaty are still unsolved.

Sovereignty remains a crucial underlying issue. Despite the apparent abandonment of CRAMRA by the Consultative Parties, its negotiation and text remain of considerable interest. The central institutions of the Convention would have been the Regulatory Committees. Each Committee would have had ten members including four claimants. Ordinary substantive decisions would demand a two-thirds majority, requiring at least one claimant to vote in favour (Article 32(3) CRAMRA). Approval of a Management Scheme[3] and of development would require that majority to include a majority of the claimant members of the Committee present and voting (Article 32(1) CRAMRA) (Auburn 1990: 261–2). For all practical purposes this could be taken to mean three of the four claimants. It may be stressed that CRAMRA has specific provisions protecting the varying positions of states on sovereignty (Article 9 CRAMRA). Therefore the voting procedures for Regulatory Committees could only be viewed as a strong practical emphasis on the continued assertion of sovereignty, as opposed to a paper formula.

Despite the continuation of concern over sovereignty, present indications suggest that the Antarctic System will continue indefinitely. The indications include the substantial number of Consultative Parties. The Consultative Parties include most of the countries capable of leading a credible challenge to the Antarctic System. A strenuous and successful effort was made to attract India, China and Brazil to become Consultative Parties, to reduce the impact of the UN General Assembly's efforts to seek a role in Antarctica (Auburn 1987: 129–32). Particular importance was attached to India because that country had sent scientific expeditions to Antarctica without becoming party to the Treaty (Auburn 1984: 401 and Dey 1991: 88–9).[4] After India's initial actions outside the Treaty System, it decided that its interests were best served within the Antarctic System. It is likely that Pakistan may reach the same conclusion for similar reasons.

113

The major criticism of the Treaty System by outside countries has come in the UN General Assembly debates on Antarctica (Auburn 1987: 125–8). The debates and resolutions of the General Assembly have been rejected by the Consultative Parties (Beck 1988: 209–10). In 1990 the General Assembly called for the full participation of the international community in the drawing up of the comprehensive convention on the conservation of Antarctica, within the UN system (Auburn 1991: 190). Once again the UN demand was rejected by the Consultative Parties. The failure of the General Assembly to gain any of its major requirements and, in particular, direct involvement in the Antarctic System, has been repeated for several years and does not seem likely to be improved upon in the future.

The final indication of the resilience of the Antarctic System is the increasing flexibility displayed by the Consultative Parties. Examples include the granting of observer status to Contracting Parties at Consultative Meetings.[5] There has been growing criticism of the Consultative Parties' record of environmental protection in Antarctica, with emphasis on specific problems such as the French airstrip at Points Geologie impacting on local fauna (Auburn 1988: 206–7) and the Argentinian vessel *Bahia Paraiso* with its consequent oil spill in 1989 (Auburn 1991). The negotiation of the Environmental Protocol to the Treaty in 1991 (*Ibid.*), despite its limitations, is an acknowledgement of the weakness of prior environmental protection measures and the readiness of the Consultative Parties to take significant steps to remedy criticisms.

IS THERE AN ANTARCTIC LEGAL SYSTEM?

Although the Treaty has been in force for thirty years, it is difficult to argue that Antarctica has a municipal legal system in any conventional sense of the term. The question of jurisdiction was specifically reserved by Article VIII of the Treaty, but attempts to raise it at Consultative Meetings have not been successful.

One can point to the extensive number of recommendations and the treaties in force (Sealing and Marine Living Resource Convention). Municipal legislation has also been passed to implement recommendations and treaties under the Antarctic System. However, such laws are limited in their application by the specific agreement being implemented and, insofar as they require jurisdiction to be exercised, apply only to nationals of the state concerned. The result is an entire continent without a domestic legal system.

CRIMINAL LAW

There is no doubt that criminal offences have been committed in Antarctica. However, most activity on the continent is directly or indirectly subject to government control. Sanctions available include the withdrawal of support for science programmes and internal discipline for naval and other military forces. The United States Navy conducts such proceedings. However, there have not apparently been conventional prosecutions for criminal activities on the continent. This may be due, in part, to the desire not to raise sovereignty issues. The question of civil and criminal law as general systems was not resolved by CRAMRA. For example, the Management Schemes were to cover 'applicable law to the extent necessary' (Article 47(q)). This may be contrasted with other international resource exploitation agreements, such as the Timor Gap Treaty (Auburn and Forbes 1991). In an interesting example of indirect prosecutions, New Zealand has charged US nationals in relation to drugs mailed from the United States to Antarctica, through New Zealand (Commander, US Naval Support Force 1988).

TORTS

The extensive litigation arising out of the Air New Zealand crash on Mt Erebus in 1979 brought out a number of important issues of Antarctic law. In *Mason* v. *Air New Zealand*, the court held that New Zealand law was applicable, after weighing the government interests of that country against the interests of the State of California. Leaving aside the interesting question of the balancing of the concerns of these two jurisdictions, one would have thought that the government interests of the USA, in the form of the basis to claim under Article IV of the Antarctic Treaty, would have been relevant to the issue of the application of laws.

New Zealand's Accident Compensation Act is a comprehensive no-fault statute eliminating accident litigation. But the Act does not generally apply to the Ross Dependency. Thus claims by the passengers' estates were filed in New Zealand and in due course settled (Auburn 1987: 177–82). The estates of the crew were covered by the Act and could therefore not sue in New Zealand. However they could and did sue the US government in the USA, alleging negligence by the Navy Air Traffic controllers at McMurdo. The action failed, but an interlocutory decision of the Court of Appeals for the District of Columbia (*Beattie* v. *US* 756F.2d.91(1984)) provided an interesting analysis of sovereignty-related issues.

115

The USA argued that since it could only be sued in tort under the Foreign Tort Claims Act (FTCA), that statute's exemption for 'any claim arising in a foreign country' (United States Code (USC) #2680(k)) applied. The Court rejected the argument by a majority of two to one. For the majority, Judge Wilkey referred to the official US government position of not recognizing any claims, but reserving its right to make claims. This suggested the interpretation that Antarctica was not a 'foreign country'. On venue, the majority held that the government position was that Antarctica has no law, so the government would say that venue exists nowhere. On the government interests test for choice of law, there were two possible systems. Antarctica has no civil law system and is not a country, so US law should apply. This analysis omits the alternative that New Zealand law might apply as that country claims sovereignty over the Ross Dependency.

Beattie could be seen as a decision on the construction of a specific United States statute where much turned on the history of the law and its purpose. However, the fact that the majority held that Antarctica is not a country and has no civil legal system appears to be part of the *ratio decidendi* of the case and of general application. The decision is not easy to reconcile with the US government official position that it is entitled to the basis to a claim. It is especially noteworthy that the US government was itself a party to *Beattie*.

In the later decision of *Smith* v. *US* (702 F.Supp. 1480(1989)), the Oregon District Court held that Antarctica is a foreign country. The case arose out of the death of an employee of the civilian contractor at McMurdo, whilst the employee was on a recreational trip off-base. 'Foreign country' refers, the judge held, to any area outside the territorial jurisdiction of the USA. The judge also invoked the argument that the FTCA requires application of the law of the place where the tort took place and Antarctica has no civil law. *Beattie* was rejected as inconsistent with prior authority.

Although the two decisions came to opposite conclusions, both relied on the argument that Antarctica has no system of civil law. In the courts of a claimant state this argument would not prevail. However, as between the Consultative Parties it is clear that there is no generally agreed system of civil law. With the increasing number of bases and personnel and the substantial number of tourists and other visitors, the likelihood of civil litigation is increased. For this reason alone it would be desirable to have a generally agreed Antarctic criminal and civil legal system, but this is unlikely because of the close relationship between jurisdiction and sovereignty.

116

AN OBJECTIVE REGIME

It has been previously argued that the Treaty did not create an objective regime opposable against all states because this would be contrary to Article IV which was intended to ensure that no such regime should be created (Auburn 1982: 117–18). However, the passage of time must raise the question of the legal status of the Antarctic System in general international law. The test of acquiescence may be applied to each new stage of the System's development (Simma 1987: 152).

Apart from the argument based on Article IV, the contentions against an objective regime still have considerable strength. Article X, requiring each Party to exert appropriate efforts to the end that no one engages in any activity in Antarctica contrary to the principles and purposes of the Treaty, suggests that the drafters of the Treaty considered third parties were not to be bound by the actual terms of the Treaty. The Treaty Regime clearly includes recommendations. There are strong arguments against recommendations binding ordinary Contracting Parties (Auburn 1982: 165–70). It is not clear whether recommendations bind new Consultative Parties. It is therefore difficult to argue that non-parties can be bound by measures under the Treaty which do not bind some categories of Treaty Parties.

This issue is a strong reason against an objective regime since recommendations provide the detailed rules of the Regime, especially on such central issues as environmental protection. The Protocol on Environmental Protection would be open to accession to ordinary Contracting Parties to the Treaty, but would not require their adherence. Since the Protocol is intended to be a major constituent of the Antarctic System, the objective regime argument is further weakened.

However, the Treaty has been in force for thirty years. For the reasons previously explained the Antarctic System is likely to be in place for the foreseeable future. The System is increasingly assuming the attributes of a permanent regime, despite its lack of a basis in conventional sovereignty. It may now be necessary to select those features of the System which are essential to its existence and to decide whether these constitute an objective regime. This task will not be easy. Several of the fundamental matters are precisely those which have aroused outside opposition, particularly in the UN General Assembly debates. One example is the right taken for themselves by the Consultative Parties to engage in a form of quasi-legislation.

117

NOTES

1 All articles in the text refer to the Antarctic Treaty or the Convention on the Regulation of Antarctic Mineral Resource Activities (CRAMRA), unless otherwise stated.
2 For example, Article 15 providing for respect for other established uses of Antarctica, including tourism.
3 In effect, the approval of exploration.
4 Pakistan's recent building of a summer station without acceding to the Treaty may well be seen in the context of its stance in the General Assembly. (See 'Pakistan builds base outside Antarctic Treaty System', *Antarctic*, 12, 4: 102 (1991)).
5 Rule 26, Rules of Procedure (1987), *Handbook of the Antarctic Treaty System* (7th ed.) (1990), 1, D4 at D5.

REFERENCES

Auburn, F.M. (1982) *Antarctic Law and Politics.*
—— (1984) 'Antarctic minerals and the Third World', *Journal of Polar Studies* 1, 2: 88–9.
—— (1987) 'Uses and exploitation of Antarctica', *Journal of Polar Studies* 4, 21.
—— (1988) 'Aspects of the Antarctic Treaty System', *Archiv des Volkerrechts*, 26.
—— (1989) 'The Erebus disaster' *German Yearbook of International Law* (GYBIL).
—— (1990) 'Convention on the regulation of Antarctic mineral resource activities' in J.F. Splettstoesser and G. Dreschhoff (eds) (1990) *Mineral Resources Potential of Antarctica*, Washington DC: American Geophysical Union.
—— (1991) 'Conservation and the Antarctic minerals regime', 9 *Ocean Yearbook.*
—— (1992) 'Dispute settlement under the Antarctic System', *Archiv des Volkerrechts*, 30.
Auburn, F.M. and Forbes, V.L. (1991) 'The Timor Gap Treaty and Law of the Sea Convention', SEAPOL Workshop, Chiang Mai, Thailand.
Beck, P. J. (1988) 'Another sterile annual ritual? The United Nations and Antarctica 1987', *Polar Record* 24.
Commander, US Naval Support Force, Antarctica (1988) *Report of Operation Deep Freeze 1987–1988*, IX–6, US Navy.
Dey, A. (1991) 'India in Antarctica: perspectives, programmes and achievements', *Polar Record* 27.
Francioni, F. and Scovazzi, T. (eds) (1987) *International Law for Antarctica*, Milan: Giuffre.
Simma, R. (1987) 'Le Traite Antarctique: Cree-t-il un Regime Objectif ou Non?' in F. Francioni and T. Scovazzi (eds) *International Law for Antarctica*, Milan: Giuffre.
Splettstoesser, J.F. and Dreschhoff G.M. (eds) (1990) *Mineral Resources Potential of Antarctica*, Washington DC: American Geophysical Union.

INDEX